IN ACTION WITH THE ENEMY

IN ACTION
WITH
THE ENEMY

The Holders of the
Conspicuous Gallantry Medal (Flying)

Alan W. Cooper

WILLIAM KIMBER · LONDON

First published in 1986 by
WILLIAM KIMBER & CO. LIMITED
100 Jermyn Street, London SW1Y 6EE

© Alan W. Cooper, 1986

ISBN 0-7183-0621-X

Photoset in North Wales by
Derek Doyle & Associates Mold, Clwyd
and printed in Great Britain by
Redwood Burn Limited, Trowbridge, Wiltshire

... Whereas Her late Majesty Queen Victoria was graciously pleased by Her Order-in-Council dated 7th July, 1874, to establish a medal designated the Conspicuous Gallantry Medal for such petty officers and seamen of the Royal Navy and non-commissioned officers and privates of the Royal Marines as distinguish themselves by acts of conspicuous gallantry in action with the enemy.

And whereas We deem it expedient to provide for the award of the Conspicuous Gallantry Medal to members of Our Military and Air Forces for acts of conspicuous gallantry whilst flying in active operations against the enemy ...

Extract from HM King George VI's approval
of the institution of the Medal,
November 1942.

Contents

List of Illustrations

9

This book is dedicated to
all Royal Air Force and Allied airmen of WWII,
whose courage is unsurpassed and
will always be remembered
as long as history is recorded.

Acknowledgements

I would like to thank all those who have helped me in research for this book; first, the holders of the Conspicuous Gallantry Medal themselves, D.J. Allen, W. Bailey, CGM, B.G. Bennett, CGM, J. Bettany, CGM, W.G. Bickley, CGM, K.W. Brown, CGM, H.A. Corbin, CGM, H.A. Donaldson, CGM, E.D. Durrans, CGM, G.F. Dove, CGM, DFM, E.F. Hicks, CGM, DFC, D.T. Jones, CGM, P. Hilton, CGM, G.F. Keen, CGM, R.B. Maxwell, DFC, CGM, J. Powell, CGM, G.E.J. Steere, CGM, W.C. Townsend, CGM, DFM, H. Vertican, DFC, CGM, J.G. Wheeler, CGM, and N.F. Williams, CGM, DFM and Bar.

In addition, I would like to thank Miss F.I. Gosling, Mrs Gosling, G.S. Richardson, W. Reid, VC, A. Saward, G. Ritchie, DFM, Allan Vile, DFC, the Dorset County Council Library Service (Miss Parker), Danford Heath Middle School (Mr Prowting), John Hayward, Diana Birch, George Griffiths, DFM, Peter Sharpe, Squadron Leader Eric Wormold, Bob Scarlett (for his great help with the photographs); Ralph Barker, the Orders and Medals Research Society (Dr W.A. Land), Air Gunners Association, Aircrew Association, the RAF Museum, Hendon, the Ministry of Defence Air Historical Branch, the Public Record Office, the Commonwealth War Graves Commission.

Thanks are also due to Hilda for her great patience in taking down dictation, to Norman Franks for his help in the presentation of this book and to the staff of William Kimber & Co Limited.

A.W.C.

Conspicuous Gallantry Medal (Flying)

CHAPTER ONE

For Conspicuous Gallantry

The Conspicuous Gallantry Medal (CGM) was first instituted during the reign of Queen Victoria, on 15th August 1855, towards the end of the Crimean War. It had been acknowledged that although the Army could award a high decoration for deeds of valour, namely the Distinguished Conduct Medal (DCM), there was no equivalent decoration for non-commissioned officers of either the Royal Navy or Royal Marines. Therefore, the Conspicuous Gallantry Medal became the Navy and Marines' equivalent.

Ten CGMs were awarded for gallantry during the Baltic and Crimean wars, but following the introduction of the Victoria Cross (VC) in 1856, no more CGMs were awarded, for acts during that conflict. The CGM decoration was not in fact awarded again until after the Ashanti War of 1873-74, in West Africa, when the CGM was re-instituted.

During the years of the First World War, the CGM was awarded to just 108 men. After World War I, with an increasing number of Royal Air Force personnel being attached for duties with the Royal Navy's Fleet Air Arm, usually at Naval Air Stations or aboard fleet aircraft carriers, an anomaly existed when World War II began. Thus provision was made, on 17th April 1940, for NCOs and men of the RAF, while serving with the fleet, also to become eligible for the award.

For the first two and a half years of the war, awards for gallantry in action for RAF pilots and aircrew were confined mainly to the Distinguished Flying Cross (DFC) for commissioned or warrant officers, or the Distinguished Flying Medal (DFM) for NCOs and occasionally airmen below the rank of sergeant. For more conspicuous acts of gallantry or for a period of sustained high endeavour in the face of the enemy, the Distinguished Service Order (DSO) was available for RAF officers, but there was no equivalent for NCOs, the DCM still being an army award.

There was, of course, the provision of the awarding of bars to both DFCs and DFMs, and naturally for an act regarded as a supreme feat of heroism, officers and NCOs alike were eligible for Britain's highest award, the Victoria Cross. Yet there was an obvious gap between DFMs (and bars) and the VC, for NCOs.

By 1942, acts of heroism within the non-commissioned ranks of the RAF, especially with the increasing number of operational aircrew in the growing Bomber Command force, were growing. The lack of suitable recognition of exceptional gallantry was put before the Air Council in the hope that an appropriate decoration could be agreed upon.

On 13th August 1942, the Air Council made two recommend-ations:

a)   that a new medal superior to the DFM should be instituted, which would allow for further recognition to airmen who had already been awarded the DFM. In the same way an officer who had been awarded the DFC could be awarded the DSO.

b)   that the DCM be awarded to airmen for gallantry on the ground in the same way as the MM which was already eligible and to correspond with the CGM and DSM when serving with the fleet.

Suggested titles for the new award for airmen were: Flying Gallantry Medal, Exceptional Flying Medal, Air Gallantry Medal, Pre-eminent Flying Medal, Meritorious Flying Medal, Conspicuous Flying Medal. The Air Council also put forward some further suggestions:

a)   that airmen should be made eligible for the DFC. (However, this would imply that the DFC and the DFM were not previously equal in merit.)

b)   that airmen should be eligible for the DSO. (This would cause difficulty with the other services.)

c)   that a Conspicuous Gallantry Medal or Star should be instituted for service in the air, awardable to both officers and airmen. (This, however, would lead to the elimination of the DSO as a high award for gallantry in the air on the part of officers. It would also extend the principle of common awards for officers and men which is not generally favoured.)

At a subsequent meeting with the Chief of Air Staff, Sir Charles Portal, and the Commanding Chiefs, they were equally divided between a medal for gallantry in the air which might be entitled 'The Distinguished Flying Star' and a medal for gallantry both in the air and on the ground, entitled possibly 'The Distinguished Service Star'. Later meetings showed the C-in-Cs were in favour of the words 'Distinguished Service'.

However, if these words were used in the title, it proved impossible to find a suitable word to complete the title. There was objections to the words Decoration, Star and Award, on technical grounds. The title Distinguished Service Medal (DSM) was already in use in the Navy.

Alternative titles were suggested: Distinguished Gallantry Medal, Conspicuous Service Medal, Distinguished Bravery Medal, Conspicuous Bravery Medal, Conspicuous or Distinguished Valour Medal.

Finally the Chiefs agreed upon the Conspicuous Gallantry Medal (Flying).

*

By Royal Warrant on 10th November 1942, the Conspicuous Gallantry Medal was extended to army and RAF personnel 'whilst flying in active operations against the enemy'. The actual medal was to be identical to the Naval CGM but was to have its own distinctive ribbon, of light blue, $1\frac{1}{2}$ inches wide with a dark blue marginal stripe $\frac{1}{8}$ of an inch wide.

Formal approval of the CGM being available for members of the RAF not serving with the Fleet was given by King George VI in November 1942. Then, on a draft Air Ministry Order dated 7th January 1943, the following order was issued:

The Conspicuous Gallantry Medal (at present only available to RAF NCOs and men serving with the Fleet) to be made available for Warrant Officers, NCOs and men of the RAF wherever serving as an award superior to the Distinguished Flying Medal, for conspicuous gallantry in air operations against the enemy. The medal would also be available for Army personnel engaged on flying duties (ie: glider pilots, observers, etc.). A very high standard of recommendation for the CGM (Flying) and DCM is to be set and maintained. Immediate awards of the CGM may be made by AOCs in Chief, of operational commands holding delegated powers of immediate awards.

*

The recommendation and approval of any award is in itself a very tenuous thing. The deed or deeds had first to be recognised by one's unit commander. This would also suppose that all such commanders would each look at an action in the same way. Clearly this is too unrealistic even to contemplate. What one commander thought was exceptional courage, another might deem to be the norm for his men. It was not unknown for acts of courage or a period of meritorious service to go unrewarded merely through a clash of personality.

Once recommended, however, the unit commander's proposal had then to be approved by a 'higher authority'. This too could meet with non-agreement for this higher authority might have his own yardstick by which he judged an award. By this same token an award might be agreed but the actual decoration changed – either way, upwards or down. The approver could downgrade or even upgrade the award. As we shall see, a few men who received the CGM were actually put forward for the Victoria Cross.

Whatever the vagaries of awarding recognition of bravery, what is contained in this book are the unique stories of 110 flying men who received the CGM. All but one received his decoration for air operations during World War II.

It is a very rare decoration: thus the men here are a very special band, yet almost unknown to the general public. When one considers that during World War II over 20,000 officers and warrant officers received DFCs and some 3,500 DFMs went to NCOs and airmen, then 110 CGMs put these men in a class of their own. Even when compared to its equivalent DSO the numbers show how rare the CGM is:

DSOs awarded in WW2

	DSO	bar	2nd bar	3rd bar
Royal Air Force	870	62	8	2
Royal Canadian Air Force	69	4	1	–
Royal Australian Air Force	69	3	–	–
Royal New Zealand Air Force	44	3	–	–
South African Air Force	34	–	–	–
Royal Indian Air Force	1	2	–	–
	1,087	74	9	2

The CGM for NCOs, came, in status, second only to the Victoria Cross for operational flying against the enemy, and in World War II only 32 air VCs were awarded. Even most air historians would be hard-pushed to name all 32 recipients, and the general public might only name one or two of the more famous, such as Guy Gibson or Leonard Cheshire. Almost none would be able to name even two CGM holders. What follows, therefore, is their very special story.

CHAPTER TWO

First Awards

It is not perhaps surprising that of the 109 CGMs awarded during World War II (one was awarded post-war), no fewer than 89 went to members of Bomber Command. Approximately 125,000 aircrew served in squadrons, OTUs and conversion units, and nearly 60 per cent of operational flyers became casualties. 85 per cent of the overall casualties were suffered on operations. Over 47,000 aircrew died in action, another 4,200 returned wounded while nearly 10,000 (many also wounded) became prisoners of war.

The honour of being the first RAF recipient of the CGM went, in fact, to a New Zealander. Flight Sergeant Leslie Bruce Wallace, aged twenty-one, from the town of Geraldine, Canterbury, on South Island, had been a railway porter before joining the RNZAF in October 1940. By late 1942 he was a wireless operator with No 83 Squadron – part of the newly created Pathfinder Force.

On the night of 21/22nd December 1942, he and his crew were on the battle order for an operation to Munich, in southern Germany. It would be a long haul for bombers during which many night fighters could be expected along the route both to and from the target. It would be Wallace's thirty-first raid, the last sixteen of which had been while 83 Squadron had become a PFF squadron.

While on the outward journey, their Lancaster (W4193 OL-A) was subjected to no less than five attacks from night fighters, the first coming from below the aircraft. This succeeded in seriously wounding the bomb aimer and set fire to some of the marker flares, as well as the camera flash in the rear of the aircraft. Another bullet hit Wallace in the leg.

Despite evasive action by the pilot, further attacks were made, wounding the mid-upper gunner. A fire was started in the waist of the Lancaster and apart from the flash and flares being set ablaze, the matting in the mid-upper's turret and everything else inflammable in the area, also caught fire and began burning furiously.

Flight Sergeant Wallace and the flight engineer went to the scene of the blaze and attempted to quell the flames. The heat, however, was very great and fumes so bad that the two men had to stop several times in order to get some fresh air into their lungs – the flight engineer was eventually coughing up blood! During the fight with the fire, Wallace was acutely aware that just below them was the bomb bay, full of incendiaries, flares and high explosive bombs. Yet ignoring his leg wound and the almost impossible conditions, he kept going until the fire was subdued and he had successfully chopped away and jettisoned all the removable burning material through the rear turret.

All the while, the burning Lancaster attracted nearby night fighters, and a number of further attacks were carried out upon them. The rear gunner was able to beat off three of these, and by skilful evasive action by the pilot, further damage was prevented. Nevertheless, the aircraft was riddled with cannon shell and machine gun hits.

When the fire had been put out and the attacks ended, the mid-upper gunner was given first aid for burns to his face and hands, then made as comfortable as possible. Wallace then returned to his wireless set and proceeded to try continuously to obtain a fix, sending messages and doing everything in his power to assist his pilot in flying their badly damaged bomber back to base. To add to their problems, the weather turned bad, with low cloud and rain over England. It was largely due to Flight Sergeant Wallace's efforts that the aircraft arrived safely back at base and it was only then that he mentioned to the medical officer that he had been hit in the leg. He was taken off to hospital to undergo treatment.

He was recommended for the Conspicuous Gallantry Medal on 28th January, by the CO of 83 Squadron, which was endorsed by Group Captain Don Bennett, the AOC of the Pathfinder Force, and finally by Air Chief Marshal Sir Arthur Harris, Commander in Chief, Bomber Command, on 2nd February. His award appeared in the *London Gazette* on 16th February 1943. Wallace was commissioned later and ended the war as a flying officer.

Recommended for the Victoria Cross

With the instigation of a new award it is perhaps not surprising to learn that there were initially actions where the unit commander recommended a higher award which was later reduced to that of a CGM. The first such recommendation was for Flight Sergeant George Ashplant, from Liverpool, a pilot flying with 166 Squadron.

Before joining 166 Squadron, Ashplant had flown with No 44 Group and on the 21st June 1942, while en route to the Middle East, had been forced through lack of fuel, to land at Portella in Portugal. He and his crew were taken to a military barracks and then to a civilian police station in Lisbon, where contact was made with the British Embassy. In a very short time they were on their way back to England, via Gibraltar, landing at Gourock in Scotland, on 12th July.

On 13th February 1943, he and his 166 Squadron crew (in Wellington BK460 AS-V) were detailed to attack Lorient naval base. It was to be Bomber Command's heaviest attack yet on this target, 466 aircraft taking part. It was to be the crew's 6th operational mission.

Following a successful attack, Ashplant was on his way home, flying at 8,000 feet above cloud, when the Wellington was struck by another aircraft that climbed up beneath them. As it crashed into the underside of the aircraft's nose, a large part of the front fuselage and fittings, including the bomb aimer's parachute stowage and 'chute, were carried away, and both propellers were smashed and sent spinning into space.

Sergeant Ashplant soon found himself out of control and losing height rapidly. After falling some 2,000 feet, he managed to regain control, gave his own parachute to the bomb aimer, then ordered the crew to abandon the aircraft. Without a parachute, he had no option but to stay with the bomber and hope to make a crash landing. This he did successfully, putting the machine down onto a ploughed field at Yeovilton, Somerset. He climbed out of the wreck to find not only the propellers but both engines had been torn off in the collision.

He was recommended for the supreme award of the Victoria

Cross on 4th March, but this was not upheld. Instead he received the CGM. Despite his selfless act of giving his parachute to his bomb aimer, no doubt the recommendation was down-graded as the episode was not caused directly by enemy action. Nevertheless, his bravery was without question.

Unhappily the luck that had been with him over southern England, deserted him on the night of 24/25th July. He and his crew, including two of the men who had been with him on the Lorient operation, were detailed for a raid on Hamburg, in Wellington HZ314. This was the first attack of what is now called the Battle of Hamburg, which was to last until 3rd August. He was on his twenty-second operation but did not return. They were shot down by flak, the eighth aircraft of twelve lost that night, shot down to crash at Buchholz. Theirs was the only Wellington lost that fateful night, and the five-man crew were buried in Hamburg cemetery.

*

The night following Ashplant's mid-air collision, the 14/15th February, saw further awards of the CGM with recommendations for VCs. The target that night was Milan, in northern Italy.

Sergeant Ivan Henry Hazard was a pilot with 101 Squadron and this night would be the occasion for a unique record in the annals of awards for gallantry. Hazard and his crew took off from Holme-on-Spalding at 6.50 pm in Lancaster ED377 SR-X. After successfully bombing the target from 11,000 feet at 10.41 pm, they were attacked by a fighter – a CR42 biplane – six minutes later. The Fiat got in a burst of fire from about 100 yards, but as it turned away was seen to be hit by return fire from the rear gunner, Sergeant Airey, and the mid-upper gunner, Flight Sergeant George Dove DFM. The Fiat went down in flames and was claimed as destroyed. In all, the gunners fired over 300 rounds between them.

The Lancaster, however, had been severely damaged. Machine gun bullets had exploded incendiaries still in the bomb bay which had failed to release and there were numerous bullet holes in the starboard centre petrol tank. The intercom had been damaged and fire had broken out in the rear part of the fuselage; then the burning incendiaries had exploded, leaving a large hole in the fuselage floor. Sergeant Leslie Airey had been hit in the legs during the attack and received facial burns. Flight Sergeant Dove recalled:

> The fighter's first burst hit the petrol tanks and wounded Airey, the rear gunner. Another set the incendiaries alight. Airey replied and set the enemy 'plane on fire. Then I gave a burst and saw it fall away blazing. While I was firing, flames and smoke rolled into my turret. My window was burned and ammunition began to explode. I scrambled down and picked Airey out of his turret, but owing to the fire and a hole blown in the bottom of the aircraft, I couldn't carry him forward.

In the meantime, Pilot Officer Moffatt, the bomb aimer, had mistaken the pilot's orders to prepare to bale out, and went out by parachute. Pilot Officer F.W. Gates, the wireless operator, Sergeant J.F.Bain, engineer, and Sergeant W.E.Williams, the navigator, all set about extinguishing the flames while Sergeant Airey, lying on the floor at the rear of the aircraft, continued to try and beat out the flames about him. Then the port engine caught fire so the pilot put the aircraft into a dive to blow it out, levelling out at 800 feet. With the rear gunner being wounded, abandoning the Lancaster was out, so Hazard would have to try and make a forced landing somewhere. By now the starboard outer engine was failing. All the escape hatches had been jettisoned in preparation for instant escape, the intercom was now totally dead and the oxygen had gone. But then Gates, Bain and Williams, succeeded in putting out the fuselage fire, and as Hazard had blown out the engine fire, he decided to try and get the aircraft and themselves home.

Hazard managed to haul the crippled bomber up to 15,500 feet to cross the alps, but then had to feather the starboard outer engine which now failed and he was compelled to make a detour and steer through the peaks rather than fly over them. The navigator, Sergeant William Ernest Williams, taking the course the pilot had been steering, worked on his dead reckoning, until he obtained an astro-fix. All his navigation was above 10/10ths cloud and his work was instrumental in Hazard getting them back to England. Williams did not receive any wireless aid until he reached the English Channel and for a period of over five hours he navigated solely by DR and astro readings. So as not to violate Swiss territory, he deliberately

(*Above*) Flight Sergeant G. Dove CGM (now commissioned)

(*Right*) Flight Sergeant L. Airey CGM

overshot his estimated time at the turning point in France by five minutes.

After leaving his position to help with the fires, Sergeant James Fortune Bain, the engineer, returned to find the starboard tank holed and leaking. He turned on the balance cocks and manipulated the petrol system throughout the return flight with the greatest skill, and on landing only some fifteen gallons of petrol were found still in the port inner tank.

Pilot Officer Frederick William Gates, the WOP, having done his share in putting out the fires, had then, with the light of a torch, rendered first aid and applied a tourniquet to Sergeant Airey's leg and given him morphine. Later he re-established intercom contact with the forward part of the aircraft by shortening the wiring, despite having to negotiate the gaping hole in the fuselage floor on several occasions. On reaching the French coast Gates returned to his set and sent out SOS signals and in making contact, was able to direct Sergeant Hazard towards the fighter field at Tangmere, in Sussex, where they landed safely in spite of having no hydraulics and only 10° of flaps.

It was only after landing that George Dove mentioned that he had been burned on the face and hands when the flames had risen to his upper turret. He had been awarded the DFM for a previous tour of operations with 10 Squadron, during which time he had flown with no less than eight different pilots.

Sergeants Bain, Airey and Williams were all recommended on 16th February for awards of the CGM, Pilot Officer Gates the DSO, while Hazard and Dove were recommended for the Victoria Cross. These two latter recommendations went as far as the AOC of No 1 Group, Bomber Command, who approved them, but upon reaching the C-in-C, were changed on 11th March to immediate awards of the CGM. All five CGMs and the DSO to Gates, were gazetted on 23rd March. A report on their Lancaster by the A.V.Roe Company, stated, 'It was the severest fire damage ever seen to one of our aircraft, and the Skipper has to be praised on his skill in getting it back.'

On returning after special leave, Sergeant Hazard was assigned a new bomber and on 20th March, he took it up on an air test. He made a low pass over Hornsea beach, but on pulling up at the end of his run, the tail wheel struck a concrete pill-box on the beach. The impact caused the Lancaster to break up. The forward section crashed into the cliffs and blew up. The tail section fell on the beach below. There were ten men aboard including Hazard, Bain and Williams. All died instantly and Hazard was buried in Oxford (Wolvercote) Cemetery.

George Dove was not with them. After the war he was told that he was wearing his CGM ribbon in the wrong place, by a ranking RAF officer. It is such a rare award that few people recognise the ribbon, and some, as the officer did on this occasion, thought it was merely a campaign ribbon and not an award for the highest gallantry.

*

A mining operation was classed as easy in Bomber Command. In fact there was no such things as an easy operation; they all entailed a certain degree of danger from either the enemy or the elements, and laying sea mines in enemy waters was no exception. For one crew such an operation, on this occasion to the Frisian Islands, had more than its share of danger and problems.

Sergeant Edward Wells Tickler was a new pilot with 49 Squadron. He had but one mission written in his log book, and that as a second pilot flying with an experienced crew. It was usual for a new pilot to fly on such a mission to gain vital experience before taking his own crew on operations. Sometimes, as in this case, it was also possible to fly a mining operation to help break in the new crew. So on the night of 27th February 1943, Sergeant Tickler undertook his first operation as captain on just such a mission.

They took off at 6.59 pm in Lancaster ED434. The first part of the trip went well, then at about 8.25 pm, flying at 1,500 feet, they were caught by the gunfire from a flak-ship. The two air gunners, Sergeants William Davies and George Bernard Silvester, returned the fire and managed to silence the ship's guns. The Lancaster, however, had been hit in the rear turret, though not seriously.

Tickler carried on but bad visibility combined with a very dark night, forced them down to 700 feet in order to make a pinpoint of their position, which was the island of Juist, which they had just located when they were engaged by searchlights. Light flak guns all opened fire at the Lancaster as it was coned by the probing fingers of light. It was hit several times in the rear turret, navigator's compartment, port wing and cockpit. The flight engineer, Sergeant Downing, received severe head injuries from which he later died. Sergeant Tickler was wounded in the shoulder and left side and for a moment, being dazed, lost control and the aircraft started to dive towards the sea. Sergeant Lowens the bomb aimer and Sergeant Silvester both yelled a warning that they were near the waves, and then the navigator, Sergeant Matthews, hearing this, left his position and went forward and, reaching over Tickler, helped to pull back the control column, bringing the aircraft back up to 700 feet.

Regaining control, Tickler made no mention of his injuries but indicated that they were going to carry on and drop their cargo. Matthews gave him a new course, and later a correct run and release was made in the correct place. It was only then that Tickler confided that he was wounded. Sergeant Matthews, too, had a slight wound on his hand and one shell splinter had torn the top of his flying helmet, though without injuring him. He gave Tickler a course for home but Tickler, now having lost the use of his left arm, could not set the course on the compass, so Sergeant Lowens had to help him and assist him to fly the aircraft.

Lowens also performed the engineer's duties and also gave Tickler tea. Sergeant Webb, the wireless operator, came forward and gave Tickler first aid, also helped with Downing, the dying flight engineer. He then went back and locked his morse key on a distress signal so that someone might pick them up and get a fix on their position in case they ditched.

It was a crew effort to fly the aircraft. All were helping each other and doing jobs they would not normally perform. Sergeant Silvester, having often watched Downing at work during air tests, took over the engineer's tasks.

After crossing the coast the wireless operator made radio contact and a flare path was lit up for them. Silvester lowered the undercarriage and with some difficulty lowered the flaps but after doing so he noticed they were losing speed too quickly and opened the throttles which seemed to help. Sergeant Tickler managed, with the help of Sergeant Davies telling him when to make the final turn into land, to make a perfect landing but he then collapsed over the controls. The Lancaster swung off to the left and struck an obstruction on the field, but little damage was done. They had landed at Doona Nook, in Lincolnshire.

On 10th March, Tickler was recommended for the DFM as were Sergeants Webb, Matthews, Lowens, Davies and Silvester. On the 17th, this was changed by Air Vice Marshal R.A.Cochrane, AOC of No 5 Group, to the following: Sergeant Tickler the CGM, Sergeants Matthews, Lowans and Silvester the DFM and Sergeants Webb and Davies, Mention in Despatches. Tickler was also put forward for a commission which he received.

On the night of 30/31st March 1944, fully recovered and now a flight lieutenant, with a new crew in 57 Squadron, Ed Tickler was detailed to fly on a raid on Nuremberg. It was an operation that would be a feature of RAF history, not for its success but for its disastrous losses – the heaviest in any one night for the whole of the war: ninety-six aircraft failed to return. Flight Lieutenant Tickler CGM, took off at 10.19 pm in Lancaster ND622. It was their thirteenth operation as a crew and proved an unlucky one.

Of the ninety-six aircraft lost, theirs was the sixty-ninth to fall, shot down by Oberleutnant Helmuth Schulte flying a Me110 of II/NJG5, who shot down twenty-five bombers over Germany during the war and won the Knight's Cross. He used on this occasion the successful method of firing upwards into the Lancaster with his Schräge Musik guns, cannons that fired obliquely upwards. Tickler, however, survived to become a prisoner at Stalag L1 at Barth, and so did two of his crew. The other four were killed.

*

Sergeant Roy Kelly Hewitt came from Newcastle and had joined the RAF in 1938. Becoming a wireless operator, Hewitt flew a tour of forty-eight ops with 101 Squadron in 1941 and received the DFM early the following year. Later that year he began a second tour with 61 Squadron. He was described as one of 3 Group's best wireless operators, and by early 1943 had reached his twenty-ninth birthday.

On the night of 10/11th March, he and his crew were to fly their last trip of their tour. For Roy Hewitt it was his sixty-ninth bombing raid – no mean total for any member of bomber aircrew. Provided Bomber Command had no 'Maximum Effort' planned, sometimes it was possible for a crew to finish a tour with an 'easy' target or operation. Hewitt and his crew were lucky. No major raids for this night were scheduled so they were sent to lay sea mines, which they did successfully.

On their way home they were suddenly attacked by four fighters on no less than three occasions. Considerable damage was inflicted on the aircraft (Lancaster W4899), flown by Flight Lieutenant Gilpin. A fire was started in the aircraft and Sergeant Hewitt assisted in putting it out and then sent distress signals by wireless. On 26th March he was recommended for a bar to his DFM but when it reached that great man of 5 Group, Air Vice Marshal Cochrane, he changed the type of reward to the CGM – an award for courage and long service covering sixty-nine bombing missions.

Hewitt had flown his last operation on a quiet night, but the next night Bomber Command sent a force of 457 aircraft to bomb Essen, where the mighty Krupps factory was situated. On this raid, Flight Sergeant Geoffrey Frank Keen was to fly his thirty-ninth op. Keen had joined the RAF in 1940 when he was twenty-four. Like Hewitt, he too was a wireless operator, had completed a tour with 51 Squadron in 1941, had been decorated with the DFM. Following a rest period, Geoffrey Keen began a second tour, this time with 427 Squadron RCAF, based at Croft, in Yorkshire.

On the night of the 12th, he was part of the CO's crew – Wing Commander D.H.Burnside DFC RAF. They took off in Wellington BK164 at 7.10 pm. The target was successfully reached at 9.30 but they were then hit by heavy flak in the target area. The navigator, Pilot Officer Heather RCAF, was killed instantly. Flight Sergeant Keen later recorded: 'I was sitting three to four feet away from him and I had my legs and feet knocked about a bit.'

In fact he had half of one of his feet blown away as well as cuts and lacerations to both legs. Despite this he regained his seat and for over two hours worked on the damaged wireless set. The Wellington had been badly hit, but the bomb aimer, Pilot Officer R.J.Hayhurst, successfully guided Dudley Burnside on the bomb run. Later, on the return flight, the rear gunner, Pilot Officer D.B.Ross, successfully beat off a night fighter attack.

Keen, meanwhile, had dragged himself to the navigator's compartment and ignoring his painful injuries managed to assist in the navigation and later obtained essential information to get them

Flight Sergeant G. Keen CGM, DFM (now commissioned)

home. They made it back to England and made a creditable forced landing at RAF Stradishall.

On 27th March, Wing Commander Burnside put forward a recommendation for Keen to receive the Victoria Cross, and who better to do so for he had been there. This was supported by the AOC of 6 Group, Air Vice Marshal G.E.Brookes CB OBE, but it was later changed to that of the CGM. Burnside himself received a bar to his DFC, while Pilot Officers Hayhurst and Ross received DFCs.

Keen was off flying for some four months but then went back and completed his second tour of operations. His tenacity in the air did not end off the ground. In civilian life he had been an avid footballer but with only half a foot he decided his football days were over: Not a bit of it. He later had a football boot adapted and he went on not just to play but to turn out for Chesham and the County of Buckinghamshire. And with, as he said recently, all with just one and a half feet! He also played cricket for Chesham, but today restricts his sporting activities to the more leisurely game of golf.

The First and the Last

Oddly, the last CGM to appear in the *London Gazette* for gallantry in World War II was for an action carried out before the actual medal was even approved. In a purely chronological sense it is, therefore, either the very first, or the very last wartime CGM.

In 1942, Paul Alexander Hilton was serving as a pilot with 35 Squadron at Linton-on-Ouse, as a sergeant. He was born in 1922 at a place called Batu, Gajah, Malaya and had joined the RAF in September 1940. When the Japanese invaded Malaya in 1941, his parents were still living there and his father was to spend the war in Changi Jail but his mother, who had been on the last boat out of Singapore, the *Vyner Brook*, died in a camp in Palembang of malnutrition and tropical diseases after the ship had been bombed and sunk off the coast of Sumatra.

After pilot training, Hilton joined 58 Squadron at Linton in October 1941 and in April '42 went on a Halifax conversion course at Marston Moor before his posting to 35 Squadron. On 2/3rd June, he was down to fly on the second Thousand Bomber raid to Essen, flying Halifax R9444 TL-D. Take-off time was 11.31 pm.

They reached Essen without incident and bombed the target, but

Warrant Officer P. Hilton CGM

on the return trip they were attacked by three Ju88s and both inner engines were knocked out. Then the attacks suddenly ceased and the crew settled down for the long trip home, on just the two outer engines. They were down to 4,000 feet when the port outer developed an internal glycol leak. Paul Hilton remembers:

I then had to give the order to bale out. The night was clear with a full moon and I thought we were partly over water. The moon shining on ground mist gave that impression. When Sergeant Prout found that he had pulled his parachute rip-cord by mistake in the aircraft, I had no hesitation in thinking I could pancake or ditch on one of the patches of, so I thought, water. It wasn't until I was just on top of the mist and I could see through it with the landing lamp, that I realised by mistake. By then it was too late. Fortunately the aircraft was about to stall and when the starboard wing hit a house the aircraft spun round through 180° and I finished up going backwards! I was not strapped in and I had a fight to open the escape hatch above my head.

I must have been unconscious for a second or two and when I came round the port engine was on fire. I managed to get clear as quickly as possible as there was still several hundred gallons of high octane close to the burning engine, so I started to run towards the cover of nearby trees. I ran between two of them and was just about to go along the road when there was a piercing scream of, '*Halt!*' from right behind me. Almost immediately I was prodded with a vicious jab in the small of my back with a rifle muzzle. I found out later I was in St Leonard, near Brecht, in Belgium, and in an area crawling with Germans. For me the war was over.

I was taken by train from Cologne to Frankfurt. The conditions were very relaxed and at one time I was actually sitting on one of the guard's revolvers which he had left on the seat. The guards were from the Luftwaffe. One was a navigator and had seen service on the Eastern Front. I was being taken to Dulag Luft, the interrogation centre at Obersal near Frankfurt. From here I was taken to Stalag Luft III at Sagan. Most of us being brought in were from the Thousand Bomber raids. I had taken part in the first one on Cologne at the end of May – my first op as captain of a Halifax.

In 1943 I was taken to Luft 6 at Heydekruge in East Prussia and then Stalag 357 at Thorn and Fallingbostel in 1944. While in Heydekruge, I stole a pair of pliers but in doing so one of the 'ferrets' (guards) heard something and asked where the pliers had gone. I then made off but was soon grabbed and hauled before the security officer, Major Peschel, and I ended up in the 'cooler'.

Hilton returned from prison camp in April 1945. He was now a warrant officer. In September he began an apprenticeship with the De Havilland engine company and stayed with them until 1955 when he joined the Austin Aircraft Company then in 1959 he formed his own engineering Laboratory Company.

It was while he was with De Havilland's that he received a letter saying he had been awarded the CGM. It had been recommended on 18th January 1946 and signed by Squadron Leader Mike Beetham of 35 Squadron, later Chief of the Air Staff. The recommendation had been countersigned by the AOC of 3 Group, Air Vice Marshal R. Harrison.

In order to attend his investiture at Buckingham Palace, he was obliged to hire an RAF uniform from Moss Bros, and in the circumstances, it seemed, as Paul remembers, to fit quite well. His award was gazetted on 29th March 1946, a most extraordinary situation for when Paul was shot down the CGM was not a flying medal and when he received it, it was for an action four years previously.

In 1984 Paul Hilton and his family took a holiday in Germany and retraced his journey when, as a prisoner, he went by train alongside the Rhine. It was just as he remembered it, even though it had been forty-two years earlier and he had been under armed guard.

CHAPTER THREE

Bomber Command – Spring 1943

Bomber Command's activities continued into the spring and early summer of 1943. As the war progressed, more and more Dominion and Commonwealth aircrew began to operate nightly over Germany and in April the first Royal Canadian Air Force recipient of the CGM was named.

Leonard Franklin Williamson, from Regina, was a sergeant pilot with 428 'Ghost' Squadron RCAF, having enlisted in 1941. On the night of 8/9th April, the target was Duisberg. 428 Squadron sent its Wellingtons out from their Yorkshire base at RAF Dalton, with Williamson in HE239 NA-Y.

When approaching the target, at 11.15 pm, the Canadian was heavily engaged by anti-aircraft fire and three minutes later the bomber was hit. The Wellington began to vibrate alarmingly and the rudder bar swung backwards and forwards, despite which, Williamson struggled to keep the machine on a straight bombing run. This was achieved and at 11.20 their bombs went down dead on target. But then, Williamson had to give the order that no pilot willingly gave – the order to abandon aircraft. However, no reply came from the rear gunner, Sergeant Lorenzo Bertrand.

Immediately the navigator went back and found the rear turret had been completely blown away and all the fuselage aft of the beam gun position, had been stripped of fabric. Williamson knew too that the hydraulics had been hit, as the bomb doors would not close and both wheels were hanging down. He also guessed that the elevators had been hit and damaged and some special signals equipment had been detonated. He was unable to climb but found he was able to maintain his present height, so cancelling the order to bale out, decided to try to get home. He reached England successfully and achieved a good landing at the fighter base at West Malling in Kent.

It had been Williamson's 7th operation and he was recommended for the DFM but this was changed to that of the CGM by the AOC of 6 Group, Air Vice Marshal Brookes on 28th April, and lastly signed by Air Chief Marshal Sir Arthur Harris on 2nd May.

The body of Sergeant Bertrand, the rear gunner, was found by the Germans and he was buried in the Reichswald War Cemetery.

A member of 488 Squadron, Royal Australian Air Force, was the next recipient of the CGM. He was not an Australian, but a Londoner; Sergeant Edward Francis Hicks had been born in Manor Park in 1912 and joined the RAF in February 1939.

He completed pilot training, mainly in Arizona in America, and then in Toronto, Canada. Here he passed as an 'above average' pilot. Returning to England in 1942 he went to AFU and then an OTU at Harwell, flying Wellingtons, which was to become his aircraft when he became operational. He joined 466 Squadron in February 1943, at Leconfield, Yorkshire.

After completing seven operations, he and his crew were detailed to fly to Stuttgart on the night of 14/15th April. In Wellington HZ256, they took off at one minute past nine, that evening. On the way to the target, at 12.40, and whilst flying at 15,000 feet, they were attacked by a twin-engined enemy night fighter.

The first indication of its presence was a burst of tracer passing over the cockpit. Hicks immediately turned to the left and dived. No contact could be made with the rear gunner, Sergeant Field, and it was later discovered that his turret had been wrecked and the gunner seriously wounded. The wireless operator Sergeant Blair, had been wounded in the leg and the bomb aimer, Pilot Officer Hopkins, and Flying Officer Clayton, the navigator, had also both been wounded.

The fighter attacked again but Sergeant Hicks took evasive action, and by turning under the enemy machine, the German pilot missed his target, and the fighter was not seen again. Sergeant Hicks then took stock of the damage. The Wellington's hydraulics and brake system had been damaged; the wheels were down and the bomb doors open, most of his crew were wounded, he had no rear turret defence, and they were still carrying their bomb load. He decided to carry on with the mission, the target being just fifteen minutes away.

(*Above, left and below*) Wellington HE239-NA-Y. (*Above right*) Seated centre, E.F. Hicks DFC, CGM (now commissioned)

They went in, and from 16,000 feet the bombs went down; then they took their bombing photograph, despite Hopkins' wounds, including a compound fracture of one arm.

For more than two hours on the return trip, the navigator, bomb aimer and wireless operator tried to get the rear gunner out of his turret. When they finally succeeded they gave him a shot of morphine but he died shortly afterwards.

When they approached the English coast, Sergeant Blair sent a message asking for assistance on landing, and on top of everything else, they were now running out of fuel. Blair made contact with RAF Ford in Sussex (today an open prison). The undercarriage was locked down by the use of the hand pump and Sergeant Hicks made a good landing without flaps. However, as the tail wheel had been damaged as well as the brakes and undercarriage, a violent swing resulted which collapsed the port wheel, but the crew were not injured any further.

Hicks was recommended for an immediate DFM but this was changed to the higher award of the CGM. Pilot Officer Hopkins was awarded the DSO, Clayton the DFC and Sergeant Blair the DFM.

Edward Hicks continued flying with 466 Squadron until August when, having been commissioned, he received the DFC on completion of his tour of ops. His tour had included missions to Düsseldorf (twice), Cologne, Aachen, Brest, St Nazaire, Kiel, Bochum and Duisberg (twice). His final award was a Mention in Despatches, on 1st January 1945. He later flew with Transport Command and finally left the RAF in 1946. In 1950, while on business in Hong Kong, he became a founder member of the Hong Kong Auxiliary Air Force.

The Dams Raid

In May 1943 occurred one of Bomber Command's most famous exploits – the raid on the German Ruhr dams by Guy Gibson's 617 Squadron. The events are described in detail in my book *The Men Who Breached the Dams* (Wm Kimber & Co, 1982).

This raid did at the time, and has since captured the imagination of young and old alike. On this operation by the 'Dambuster' Squadron, as they became known, two men won the CGM.

At the time of the raid, Flight Sergeant William Clifford

Townsend was twenty-two years old. He had joined the army in 1941, but soon transferred to the RAF. When he joined 617 in March 1943, he had already completed a bomber tour with 49 Squadron. On the Dams Raid on 16th May, he was detailed as part of the reserve force, which if necessary, would be called upon to support or attack other dams after the Möhne and Eder Dams had been destroyed.

Townsend took off at 12.14 am in Lancaster 'O' for Orange (ED886) and was ordered to attack the Ennepe Dam on the Schelme River. He made three runs on the dam before his bomb aimer, Sergeant Charles Franklin DFM, was satisfied. The bouncing bomb, invented by Barnes Wallis, was released at 3.37 am. It bounced once and exploded thirty seconds after release. On leaving the target much opposition was encountered but by great determination on his part, plus the navigational skill of Pilot Officer Cecil Howard, from Australia, they made it safely back to their base at RAF Scampton, landing at 6.15 am. Most of the latter part of the homeward trip was flown in broad daylight.

The second CGM awarded went to Flight Sergeant Kenneth William Brown. He was twenty-three years of age, hailed from Canada and had enlisted in the RCAF in 1941. He had also been part of the reserve force and took off in Lancaster 'F' for Freddie (ED918) at 12.12 am. His trip was quite eventful. Even before he reached the Sorpe Dam, which he was assigned to attack, his gunners shot up three trains en route. He was fired upon by flak and hit in the fuselage but suffered no serious damage to the aircraft.

He made eight runs on the Sorpe, but was still not happy about dropping his bomb. On the ninth he dropped incendiaries on the banks of the lake to try and identify the dam through the swirling mist. On the eleventh run they saw the dam and his bomb was released. This was at 3.23 am. The bomb exploded on impact, as it was dropped while flying across the dam and not dropped towards it in the planned method, without it bouncing off the surface of the water. A crumbling of some 300 feet of the dam wall was seen, the dam having already been damaged by an earlier attack by Flight Lieutenant Joe McCarthy.

Ken Brown made it safely home again, landing at Scampton at 5.33. Both awards were recommended on 20 May, three days after

Take-off for the Dams (by Maurice Gardner)

(*Left*) Flight Sergeant Bill Townsend CGM, DFM (now commissioned) (*Right*) Flight
Sergeant Ken Brown CGM (now commissioned)

Ken Brown in 1983

Townsend returns (by Maurice Gardner)

the raid. Eleven of the nineteen Lancasters had failed to return but the force had destroyed the Möhne and Eder Dams, and damaged the Sorpe. Guy Gibson was awarded the Victoria Cross for his leadership and in all, 617 Squadron received six DSOs, ten DFCs, six bars to the DFC, eleven DFMs and one bar to the DFM, as well as these two CGMs – a total of thirty-seven gallantry awards.

*

In most cases the pilot, like the captain of a ship is, traditionally, the last to leave the aircraft. If he orders his crew to bale out, it is he who has to hold the aircraft steady in order for them to get out safely. Yet there were instances of other members of the crew flying the aircraft home, either because the pilot had been killed or wounded, or as in the case of Wellington HE198 SE-D, where the pilot had baled out.

The target for HE198 of 431 'Iroquois' Squadron RCAF, on the night of 23/24th May 1943, was Dortmund. A total of 826 bombers flew on this raid, the greatest number so far with the exception of the Thousand Bomber raids a year earlier. They took off at 10.40 pm and bombed the target from 17,000 feet. Only seven minutes later HE198 was coned by searchlights. The pilot immediately put the aircraft into a steep dive, but soon discovered that he was unable to pull out and called for assistance. This came from Sergeant Stewart Nimmo Sloan, the bomb aimer. He put his back to the instrument panel and pushed while the pilot heaved back on the column. This effort worked and they put the Wellington onto level flight. They were still coned, however, and also hit by heavy flak. The pilot put the aircraft into another dive when the rear gunner, Sergeant Warne, shouted that they were on fire. On hearing this, the pilot ordered everyone out and then baled out himself. There appears to have been some confusion as to the actual order and Sloan quickly realised that not all the crew were yet out and decided to try and pull the bomber out of the dive it was in. He succeeded and when the navigator, Sergeant Parslow, came forward, he found Sloan sitting sideways at the controls, being prevented from sitting correctly because of his parachute. He had clipped on his chest type 'chute when the pilot first mentioned baling out.

They then found they were no longer on fire and as the

Wellington was still answering to the controls, Stewart Sloan decided to try and fly it back to England, despite the searchlights and exploding flak shells. At that moment the emergency call light flashed and he called for his flying helmet which he had previously discarded. Putting it on he began to call up the crew positions to see who still remained on board. He found that Sergeant Warne had already jumped, but the WOP, Flying Officer Bailey, and the navigator were still in the aircraft.

He then tried to get out of the searchlight's glare which he did at 4,000 feet, but then they picked them up again, but at least this time the flak had ceased. Suddenly the lights lost them. They later estimated that they had been coned for some fifty minutes in all.

It was now time to regain some of the lost height and set a course for base. Sloan got it up to 9,000 feet and headed out on a rough course, helped and supported by Bailey and Parslow. During the evasive action the emergency escape hatch had been blown open. The downward exit could not be closed and with the rear turret doors hanging open, a tremendous gale was blowing through the length of the bomber. To add to Sloan's problems, the revs on the port engine started to rise, caused by the failure of the engine drive generator. Sloan placed the propeller in fixed pitch and gave orders for all unnecessary services to be cut out.

He headed towards the nearest point of the English coast – Orfordness. This he reached successfully then headed towards their base at Burn, via the only landmark beacon that he knew, that of Cottesmore. However, fixing the pitch of the port propeller had not been entirely successful and he had to throttle back to keep the revs down. The Wellington was now down to 3,500 feet and Sloan was having difficulty in even maintaining this height. He took the decision to land at the first aerodrome showing the Drem lighting system. However, the first field he saw lit was a simple one and so he decided he had to try and get down. He circled three times and they fired a red Very cartridge. In reply the airfield fired a green and in he came, the starboard engine cutting out as he did so. His landing was perfect. The airfield turned out to be RAF Cranwell, the air force's own college, in Lincolnshire.

His flight back and landing were remarkable as Sloan had no piloting experience except for the few occasions he had taken over

(*Above*) Sergeant Sloan being congratulated

(*Right*) Sergeant Sloan CGM (now commissioned)

the controls under the watchful eye of a pilot. He learnt most of what he knew by watching others and by constant practice on the ground Link Trainer.

Sloan was recommended for the CGM on the 24th, given an immediate commission in the field (very rare in World War II) and he accepted a place on the first available pilot's course. Sergeant G.C.W.Parslow was awarded the DFM and Flying Officer J.B.G.Bailey the DFC. Both Bailey and Parslow became part of the crew of Wing Commander J.Coverdale, the squadron CO, but were killed on a raid to Krefeld on 21/22nd June. Sloan, later a flight lieutenant, returned to operations as a pilot with 158 Squadron in January 1945, winning the DFC. After the war he flew with the King's Flight, retiring in 1951 with the MVO (Member of the Victorian Order).

*

On the very day that Stewart Sloan's CGM appeared in the *London Gazette*, another airman was about to win this coveted award.

Flight Sergeant Norman Francis Williams was an Australian from Leeton, New South Wales, where he had been born on 3rd November 1914. He enlisted in the RAAF in May 1941 and became an air gunner in March 1942 with the rank of sergeant. He was trained in both Australia and England before being posted to 10 Squadron in August 1942. On his tenth operation, a raid upon Bremen on 5th September, and while returning from what till then had been a straightforward raid, they were at 8,000 feet over the Dutch coast when Williams spotted a Ju88 approaching from the port quarter. It passed below, turned on track and followed them eight feet below their tail, but out of Williams' view from his rear turret. The German pilot was an experienced night fighter and knew how to find a bomber's blind spot.

Williams called for a steep climb and his pilot responded immediately and everyone else in the crew left their stomachs behind, then the fighter came into view. Williams opened fire with a long burst and saw the fighter burst into flames, illuminating the sky, before plunging to the ground. This was seen and confirmed by other members of the crew and also by another crew from a squadron based at Topcliffe, who was flying four miles away. For this action Williams was awarded an immediate DFM.

On 30th March 1943, he was recommended for a bar to his DFM following the completion of his tour of ops. He had flown thirty in all, including attacks on Berlin, Cologne, Turin, Genoa and Hamburg. By this time he had already volunteered to be posted to 35 Squadron at Gravely, a squadron of the Pathfinder Force. By now he was a flight sergeant and on the night of 11th June, and with a very creditable total of forty ops under his belt, he and his crew set off for Düsseldorf as rear gunner in Halifax HR798 TL-A, flown by Pilot Officer Cobb.

On the outward journey and nearing the target area, they were attacked by two night fighters on three occasions. In the first attack both Williams and the mid-upper gunner, Flight Sergeant Smith, were wounded. Smith was grazed on the head by a bullet and had complete loss of memory. Williams was hit by bullets in the legs and one near to his stomach. In the second attack Williams opened fire with a long burst of about 300 rounds. The attacking fighter ceased firing and exploded, pieces being seen falling in flames. In the third attack Williams again opened fire with 300 rounds. The fighter started to dive down with pieces coming from it but he followed it down, still firing, with guns fully depressed. The flight engineer saw large lumps of the fighter floating down until it disappeared into clouds.

With the aircraft damaged and both gunners wounded, Cobb ordered the bombs to be dropped on the outskirts of Düsseldorf, and then he set off for England, Williams refusing to leave his turret. When they finally made it home, his turret was found to be so badly damaged that he had to be cut out of it. He was recommended for the CGM four days later.

His wounds kept him out of action until August, when he returned to 35 Squadron where he remained until January 1944. He was commissioned in October 1943 and ended the war as a flight lieutenant. He returned to Australia in July 1944 but in February 1945 was posted to 23 Squadron RAAF, with whom he served in the Pacific, operating from Morotai in the Helena Heras Islands. He carried out 36 operations against the Japanese in Borneo and Japanese occupied islands. He recalls: 'These trips were a milk run compared with those over Germany – enemy opposition was negligible.'

After the war he was with the Occupational Forces in Japan until

(*Above left*) Bust of Warrant Officer Norman Williams CGM, DFM (now commissioned)

(*Above right*) Turret Door of Norman Williams' aircraft

(*Left*) David Codd and Norman Williams 1985

1948 when he returned to Australia, to be discharged in May, with the rank of squadron leader, and one of the most decorated air gunners of World War II. He rejoined the RAAF in 1951, flying Lincolns in RAAF Transport Command in Singapore and in 1953 was posted to Korea as operations officer for 77 Squadron RAAF. He was demobbed in 1954 and settled down to farming in New South Wales. Now 72, he is still farming and despite losing the sight of one eye is still a crack shot. His damaged gun turret now resides in the magnificent War Museum in Canberra, Australia.

The month of June 1943 was to see further acts of bravery that resulted in a number of aircrew being awarded the CGM. The first of these occurred on the night of 22/23rd and it went to another Australian.

Bomber Command was fighting the Battle of the Ruhr which commenced on 5/6th March and lasted through till 24th July. On the night of the 22nd, 557 bombers went to Mulheim. One pilot detailed for this raid was Flight Sergeant Francis Edwin Mathers, serving with 77 Squadron. He had been born on 8th November 1921 and enlisted in the RAAF in August 1941. After training in Australia he came to England in September 1942 and after completing his training at an OTU and Conversion Unit, was sent to 77 Squadron at Elvington, Yorkshire, on 3rd May.

The trip to Mulheim was his eighth. The flight out, in Halifax JD110 KN-P, was uneventful and he and his crew bombed the target at 1.48 am from 19,000 feet. As the bombs were released the aircraft was hit by heavy flak and the starboard outer engine caught fire. This was feathered and the fire subsided. Three minutes later they were again hit, this time causing the port inner to burst into flames, and this too was feathered. The damage to the Halifax was considerable. Besides the two engines being knocked out, the starboard aileron control was non-existent and three petrol tanks had been holed, two of which immediately emptied while the other began leaking. Mathers headed for home but was unable to maintain height.

On reaching the North Sea they were down to 3,000 feet and as soon as the Halifax was clear of the enemy coast, Mathers ordered all non-secret equipment that could be removed to be jettisoned, followed by the guns and ammunition from the upper turret. When down to 2,000 feet and well out to sea, they were confronted by a Me110 which made three determined attacks. The first destroyed communications with the rear gunner, Sergeant Speedie, and the ammunition tracks to the rear turret were damaged. Mathers turned the aircraft into a slight diving turn which enabled Speedie to fire two short bursts before he had a stoppage.

While Speedie cleared his guns, the fighter came in again and opened fire from 400 yards but Speedie was back in action and got in a short burst. By now they were down to 1,000 feet and the 110 was coming in for a third attack. Speedie opened fire from 200 yards and the 110 dived into the sea. The altimeter needle now rested on 500 feet and when they reached the English coast it had unwound to 400. The WOP, Sergeant George French, was instructed by Mathers to obtain a MF/DF fix which he did and also repaired a damaged trailing aerial which had earlier been shot away. They were also flying on almost empty fuel tanks.

They reached Martlesham Heath, usually a fighter airfield, where Mathers found his wheels would not come down, so he had no choice but to make a belly landing. He brought the crippled Halifax down with the minimum of additional damage and without injury to any of the crew. For the final fifteen minutes the fuel tanks had been reading zero!

Sergeant French at his radio, Speedie at his guns and the superb flying skill of Francis Mathers had got them home. On 26th June, they were all put up for awards, the CGM for Mathers, DFMs for French and Speedie.

On 5th September, with the same crew plus Sergeant K. Adams, flying a second pilot trip, Mathers took off in Halifax JB839 – 'K' to attack Mannheim, and failed to return. Their aircraft was found the next day. Five of the crew, including Adams, Speedie, French and the gallant Mathers, were dead. Although shown as a flight sergeant on the battle order, records show him as being a pilot officer on this date. Commissions were often late in coming through but when they did were usually dated from the date of the interview. Mathers was twenty-one when he died and was buried in the Durnbach War Cemetery in Germany, alongside French and Speedie.

Less than a week later, 28/29th June, following Mathers' trip to

Mulheim, Bomber Command went to Cologne with a force of 608 bombers. Piloting one was 22-year-old Sergeant Cecil James Morley Wilkie, from Catford in south-east London, who had already flown fifteen missions. He took off in Lancaster ED753 at 11.04 pm, from his base at Skellingthorpe, the home of 50 Squadron.

Over Cologne and while on the bombing run at 19,500 feet, the aircraft was coned by searchlights and hit by flak. The cockpit perspex was smashed and Wilkie blinded by dust and splinters. The same flak hit wounded the bomb aimer and navigator, but despite this, the run was continued and Pilot Officer Hearn, the wounded bomb aimer, keeping the knowledge of his wound to himself, guided the half blinded Wilkie over the target and released the bombs.

Turning for home, they were under shellfire from Cologne to Düsseldorf and it was only then that Wilkie was able to clear the sight of his left eye. He then managed to evade probing searchlights and set course for England and home. To add to his difficulties, the intercom had also been knocked out.

The nose of the Lancaster was riddled with heavy and light flak. Hearn in fact had been hit in three places in his left arm, and the navigator, Sergeant Heath, wounded in the right shoulder by shrapnel. The second navigator, Sergeant Forts, had also been cut about the face by flying perspex. Wilkie, still unable to see clearly, made it back to base and landed safely at 3.58 am. He was awarded the CGM while Hearn received the DFC. Hearn was later to earn a bar to his DFC for successfully evading capture after being shot down. Sergeant Wilkinson, another member of Wilkie's crew also received the DFM.

Wilkie was later commissioned and continued with his tour but like Mathers was fated to die in action. He was detailed to bomb Frankfurt in Lancaster W4905 VN-H, taking off at 6.30 pm. They failed to return, falling in the southern suburbs of Frankfurt-am-Main. He was killed as were two other members of his crew. He is buried in the Durnbach War Cemetery.

*

In another bomber over Cologne on the night of 28/29th June was Sergeant Edwin Thomas George Hall of 115 Squadron. On this night he was the mid-upper gunner in Lancaster DS669.

At 2.42 am on the homeward run, having made a successful attack

(*Left*) Flight Sergeant F.E. Mathers CGM

(*Below*) Sergeant E.R.G. Hall CGM (3rd from left) and crew at Buckingham Palace

on the target the rear gunner, Sergeant White, reported two yellow flares at the same height as themselves. He then reported two single-seater FW190 'Wild Boar' fighters on the port quarter and slightly below. The pilot of the Lancaster, Sergeant Jolly, corkscrewed the bomber to port and both gunners opened fire at 250 yards. One of the Focke Wulfs broke away without opening fire, but the other attacked and fired a short burst of both cannon and machine gun fire that passed behind the Lancaster. White managed to get in another burst and the fighter broke away and was not seen again.

The first 190, however, was not put off quite so easily. It came back and attacked from 200 yards from dead astern. As the Lancaster corkscrewed to starboard the gunners opened fire but so did the fighter and the rear turret was hit, killing Sergeant White, and setting the fuselage on fire behind the bulkhead doors. The fighter kept on attacking down to 50 yards. Despite the burning fuselage beneath him, Edwin Hall stuck to his two guns and kept firing back until the 190 broke away below them and to starboard. It was last seen by the bomb aimer, Sergeant Gourley, on fire and in a steep dive – it was claimed as 'probably destroyed'.

At this point Hall was obliged, because of the intense smoke and flames, to leave his turret. The WOP, Sergeant Crowther, with a portable fire extinguisher, was fighting the flames and helped by Hall, managed to put out the fire. Only then was it discovered that in the last attack the rear turret had been shot completely away, taking the body of Sergeant White with it. He had been twenty-one and came from Thornton-le-Dale, Yorkshire. His body was never found but his name is recorded on the RAF Memorial at Runnymede.

Sergeants Hall and Crowther were both exhausted by this time as there had been no chance to connect up to a portable oxygen supply. With the fire out, Hall returned to his turret and Sergeant Jolly managed to fly home and make a successful landing even though the tail trimmer had also been damaged. Both propellers on the port engines had been damaged by cannon fire and the intercom knocked out. Hall was awarded the CGM, and Jolly and Crowther were deservedly rewarded with immediate DFMs.

CHAPTER FOUR

High Endeavour

Like a number of other decorations, the CGM, as well as being awarded for specific feats of bravery, could also be a reward for a number of actions or for a period of continued gallantry of a high order. During 1943, there were ten men of Bomber Command who were thus honoured.

The first two came from the same squadron, 156, flying Avro Lancasters from RAF Warboys, Huntingdonshire, flying as part of the Pathfinder Force.

Flight Sergeant Anthony Elcoate had been awarded the DFM in February 1941 having flown on thirty operations with 10 Squadron, flying as a WOP/AG, in twin-engined Armstrong Whitworth Whitley bombers from RAF Leeming, Yorkshire. Two years later, on 21st February 1943, and now with 156 Squadron, he was recommended for the CGM having completed a total of sixty-two bombing ops. In that time he had trained himself as a bomb aimer although his aircrew category remained as a wireless operator/air gunner.

He had flown against all the tough targets such as Kiel, Bremen, Duisberg, Stuttgart, Munich, Düsseldorf, as well as to Turin and Milan in Italy. His very first bombing mission had been flown as early as 5th July 1940.

Warrant Officer Forbes McPherson Taylor, had completed fifty operations by early 1943. He had been born in Calcutta in 1920 and had enlisted in the RAF in 1939.

He was recommended for the CGM on 21st February, having flown against major targets in Germany and Italy as a pilot. This was gazetted on 20th April. On the day following the publication of his award, he was recommended for the DFM having brought his total ops to fifty-nine. By June he had been commissioned.

*

Flight Sergeant Robert Service Hogg was born in Berwick on Tweed in 1922, and enlisted in the RAF in 1938. Having been an electrician in civilian life it was natural that he should become a wireless operator/air gunner. He was put forward for the DFM in March 1943 but, as was the prerogative of senior officers, Air Vice Marshal Ralph Cochrane, 5 Group's AOC, had it changed to that of the CGM. At the time Hogg was flying with 49 Squadron on Lancaster bombers at RAF Scampton and then Fiskerton, and had completed sixty-four operations.

Sergeant Edward Ernest De Joux was serving with 102 Squadron in 1943. He had previously flown with 7 Squadron in 1941-42 and in mid-December '41 he was recommended for a DFM following his eighteenth operation, a raid upon Brest. It had been a daylight mission and his aircraft was attacked by two Messerschmitt 109 fighters. In their first attack the hydraulic mechanism was put out of action, but by skilful use of his turret hand rotation gear, he managed to shoot down the second of the two 109s which fell in flames. The Stirling bomber in which he was flying was damaged but they made it home safely.

He then went to 102 Squadron and flew a further nineteen ops between January and May 1943, leaving the squadron in June. At the end of his tour he was put forward for the CGM although the actual recommendation seems not to have survived, but the award was gazetted on 12th November of that year.

The CGM and DSO, plus the DFC

Warrant Officer Bernard William Clayton, known as 'Bunny' to his friends, was awarded a DFC in April 1943. By the time this had come through, he had been recommended for a bar to this decoration, both awarded while flying with 51 Squadron, flying Halifax bombers from RAF Dishforth and Snaith. However, on 30th April the award of the second DFC was changed by Air Vice Marshal C.R. Carr, AOC of 4 Group, to that of the CGM. Clayton had completed fifty-one operations starting with his first on 9th May 1941 and ending with one to Pilson on 16th April 1943.

In 1944 he commenced another tour as a flight lieutenant with the famous 617 'Dambuster' Squadron. He was awarded the DSO (which is the equivalent award for officers as the CGM is for NCOs). He now had flown seventy-seven operations and with 617 he had taken part in many of their tough raids, such as against the Anthéor Viaduct, Watten and Wizernes (the V-Weapon site targets), the E-boat pens at Boulogne. The last operation, mentioned in his recommendation, was against Siracourt on 26th June 1944.

*

On the night of 14/15th January 1944, Warrant Officer Ronald Haywood of 7 Squadron of the Pathfinder Force was reported missing from a raid on Brunswick. His Lancaster crashed at Querium, near Brunswick, and he is buried in the Hannover War Cemetery, in Germany.

He was born in 1921 in Staffordshire and had enlisted in 1939. When he was recommended for the CGM in November 1943, he had flown eighty-three operations, fifteen of which were as part of the Pathfinders, six as marker crew. He flew as a wireless operator and his early missions were flown in the Middle East. His CGM was gazetted on 18th January 1944 – three days after he had been reported as missing in action.

Only a month after being awarded his CGM, Warrant Officer James Mitchie Alexander, also of 7 Squadron, failed to return from Berlin. It was the last but one raid of the Battle of Berlin series, and he was flying as navigator to Flight Lieutenant P.K.B.Williams in Lancaster ND365 MG-L. They were attacked by a night fighter and all but two of the crew were killed, including Alexander. He is buried in Skaro Cemetery, on the Isle of Skaro, Denmark.

He had been born in Wood Green, London, in 1920 and enlisted in 1939. Like Haywood, his first operations were flown in the Middle East. After sixty such operations he returned to England and joined 7 Squadron and had flown a further nine raids with the Pathfinders when he was recommended for the CGM in November 1943. His loss on 15/16th February 1944 was another blow to his squadron.

Pilot Officer B.W. Clayton DSO, CGM, DFC, Flight Lieutenants Maltby and Martin

With four squadrons

On the same date as Bunny Clayton's CGM award appeared in the *London Gazette*, Flight Sergeant Ivor Ward Preece's name also appeared as a recipient of this award. He had a varied service as a rear gunner, having flown fifty-seven ops by the time of his award in June 1943. He had flown with 9 Squadron in England, then with 38 Squadron in the Middle East, both on Wellingtons, taking part in the Western Desert campaign. He had then returned to England and flew with Guy Gibson's 106 Squadron on Lancasters. He was known as 'Taffy' but came from Ireland.

Preece then went to 83 Squadron and flew as rear gunner in the Lancaster flown by Group Captain John Searby, CO of 83 Squadron and leader of the famous raid on the Peenemünde rocket development establishment, on 17/18th August 1943. The raid was the first in which a Master Bomber controlled a full scale raid. Guy Gibson had been first to use the method during the Dams Raid but that had only been a small scale raid. Peenemünde was attacked by a force of 596 bombers, rather more than Gibson's nineteen Lancasters.

On the Peenemünde raid, Searby's aircraft was attacked by a Me110, five miles from the target on the homeward trip. Both Preece and the mid-upper, Flying Officer Coley, opened fire and the fighter was seen to turn to starboard and come up on the starboard bow above them. It then proceeded to turn in front of their Lancaster and disappear below. It did not return the gunner's fire. He was still flying ops in December, in Flight Lieutenant W. Thompson's Blind Marker Crew.

Another flyer who was lost soon after receiving the CGM was Warrant Officer Clive Camden Busby of 156 Squadron, operating with Avro Lancasters. He failed to return from a mission on 16th June 1943 and is buried in the Rheinberg War Cemetery. He was twenty-two years old and his parents came from Ceylon.

He was recommended for his CGM on 21st May 1943, after completing fifty-three raids against all the major targets, the last five while being with the Pathfinder Force. Sadly he was dead before his award was gazetted.

On 28th June 1943, Warrant Officer Michael George Clynes had forty-seven operations in his flying log book when he was put forward for a CGM. He was an air gunner with 431 'Iroquois' Squadron, RCAF. Forty-five of his operations had been flown with 104 Squadron in the Middle East flying Wellingtons, over Egypt and Libya in North Africa. Over Europe he flew against Berlin, the Ruhr and over Brest. During his operations there were many interesting comments noted in his log book such as: 'shot-up and crash-landed', 'harness cut by shrapnel', 'attacked by a Me110', 'cannon shell through rear turret', 'bombed from 10,000 feet', 'six holes in the aircraft', 'four thousand pound bomb hung-up, chopped free with an axe after three runs over the target' (the target was Benghazi in North Africa), 'supply dropping in Crete', 'shot down a Me109 – badly shot-up'.

Mike Clynes shot down two certain enemy fighters and aircraft in which he had been flying had crash landed on no less than four occasions. His CGM was approved and gazetted in August 1943. On 25th November he took off as rear gunner to Pilot Officer Horton in Halifax LK975 SE-P, and failed to return. He is now buried in the

Durnbach War Cemetery which is the nearest in the south of Germany, to Frankfurt, which was the target that night. When he was killed he had fifty-seven operations noted in his log.

CHAPTER FIVE

The Men of Coastal Command

Coastal Command aircrews also received their share of CGM awards during 1943-44, although the lion's share did go to Bomber Command. The first of Coastal's CGMs went to Flight Sergeant Charles Clayton Corder who was born in 1917 at West Thurrock, Essex.

He was turned down for aircrew duties on enlisting in the RAF, as his left eye was slightly below normal vision but after he had taken a wireless operator's course at Yatesbury, he was sent to an Initial Training Wing at Torquay for a navigation course. Passing out as a navigator, he joined 248 Squadron flying Beaufighters, in November 1941, and crewed up with a Lieutenant Max Geudj DFC, a Frenchman of the Jewish faith. To safeguard his family in France, Geudj was known as Lieutenant 'Maurice' in the RAF, and with him, Corder flew seventy-five operational sorties.

On 10th March 1943, in Beaufighter 'W', they took off for a line patrol, in the Bay of Biscay. They encountered and shot down a Ju88 long range fighter although Geudj was slightly wounded and the aircraft severely damaged by the return fire from the Junkers. The intercom was put out of action so Corder came forward and stood behind his pilot, worked out an accurate course for base, then went back to try and contact the other Beaufighters with an Aldis lamp. He wanted to tell them to escort them back to base but the lamp too was out of action. He again went forward to help Geudj, pulling back the jammed port airscrew pitch control lever at the same time.

As he attempted to get the lever back it came away in his hand. He went back to his radio and sent a message to 19 Group HQ that they were returning to base and that Geudj was wounded. He then repaired the intercom but then became aware that the port engine had failed and that the aircraft was losing height. Then the good engine caught fire and another fire of burning oil broke out in the cockpit. Corder sent out an SOS as he thought they must soon ditch but then the fires went out so they staggered on. They managed to reach the airfield at Predannack where Geudj put the Beau down in a crash landing. The machine burst into flames and on escaping out of the top hatch, Corder ran round to see if Geudj was safe, which fortunately he was. The journey home had covered some 180 miles.

Corder was recommended for the CGM while Max Geudj received the DSO. Geudj went on to become CO of 143 Squadron as a wing commander but was lost on an operation to Norway in January 1945 flying a Mosquito. After the war Charles Corder returned to his pre-war occupation in banking.

A month after Corder's decoration came the first CGM awarded in the RAF's anti U-boat war. Sergeant Arthur Francis Blackwell was born in London on 10th August 1919. He was living in Australia when war began, so enlisted in the RAAF in 1941. Most of his training was done in Canada and then England, before joining 500 Squadron of Coastal Command in July 1942, as a navigator.

The squadron was based at Gibraltar, but in the spring of 1943 had a detachment at Blida in North Africa, flying anti-submarine sorties out over the Mediterranean. On 23rd April, Blackwell was flying with his pilot, Warrant Officer Obee (who the previous November had shot down a Ju52) in Hudson 'N'. They spotted a U-boat on the surface and Obee made an immediate attack with depth charges, from 200 feet. However, when they were still 200 yards from the German submarine, it opened fire with cannon. The first burst passed over the starboard engine, but then a shell burst inside the cockpit, hitting Obee in the stomach and setting his clothes alight. The crew later discovered that Obee must have died instantly. With the pilot apparently unconscious, Flight Sergeant Kempster controlled the Hudson while his body was taken from the seat, and Blackwell closed the bomb doors.

Arthur Blackwell then took over the controls and flew the Hudson back to base. When above the airfield, Blackwell ordered the crew to bale out as he was not certain he could land the Hudson safely, and they knew now that Obee was dead. After the crew baled out, Blackwell headed the aircraft out to sea then baled out himself, the

(*Left*) Flight Sergeant C.C. Corder CGM

(*Below*) Hudson on patrol

Hudson crashing later on a salt bank. He was recommended for the CGM on 7th May, and Kempster received the DFM.

On leaving 500 Squadron in September 1943, Blackwell returned to England and was later commissioned. In March 1945 he served with 466 Squadron RAAF in Bomber Command, flying Halifaxes until the war's end. He returned to Australia in July 1945 and was demobbed in October with the rank of flight lieutenant.

*

Coastal's task of seeking out the elusive German submarines often entailed hours and hours of boring sea patrols. Flying out over the Atlantic, the Western Approaches, the northern transit area around the north of Scotland, or across the North Sea to the Norwegian coast, they had to fight the elements as well as the known enemy. Another area which became deadly, not only for U-boats but also for Coastal Command aircraft, was the Bay of Biscay. German submarines had to cross this expanse of water to leave or to return to their French bases, so it was constantly patrolled.

By 16th May 1943, Sergeant James Stephen Powell of 224 Squadron had carried out twenty-six anti-submarine patrols. On the afternoon of this day he piloted his huge B24 Liberator (FL948 'M') over the Bay and spotted through binoculars a submarine on the surface. The conning tower he could see had two stepped platforms at the rear on which were mounted machine-guns but only one deck gun forward of the conning tower.

He attacked from the U-boat's port beam, coming in low and fast. Usually U-boats tended to crash dive when seen, but their captains had recently been ordered to stay on the surface if surprised and fight back. Jim Powell recalls: 'It was the first action any of us had where the U-boat actually stayed on the surface and indulged in a shooting match.'

The U-boat opened fire with the heavy calibre deck gun – estimated 4-inch – and also with machine guns from the conning tower. This was totally unexpected and the Liberator received damage. Jim Powell had his hand on the button ready to release the six Torpex-filled depth charges, when a shell exploded right in front of him, damaging the Lib's port inner propeller and blowing the nose wheel doors inwards. The explosion caused Powell to press the button and release the D/Cs prematurely. As they flew over the boat, the rear gunner estimated that the nearest D/C fell thirty feet astern of the submarine. He opened fire as they passed over, estimating hits on the boat, and one of the three or four men seen on the stepped platform, fell, possibly overboard. Sergeant Powell turned to port to attack with his second stick of D/Cs but the U-boat was seen to be diving. A second attack was made down the U-boat's track, Powell releasing six D/Cs from eighty feet. He aimed just short of the swirl of the diving submarine as it went below. The rear gunner again saw one explosion short of the swirl and a final three astern of it. The mid-upper fired a short burst as they flew over. Only some diesel oil and scum was to be seen when they circled the spot where the enemy vessel had gone down. After fifteen minutes they resumed their patrol although they had now used all their D/Cs.

Jim Powell had suffered concussion and the second pilot now had to fly the aircraft for about thirty minutes before he was able to take over again. During the next hour or so they saw two further U-boats and made simulated attacks, forcing them to submerge: 'We made a dummy run and scored many hits with our ·50s and ·303s, forcing the two submarines to crash dive but on these occasions, having dropped our depth charges on the first submarine, we had no more to drop on these two.'

For his courage and determination, Jim Powell was awarded the CGM. He continued to fly with 224 Squadron, but after this momentous day he was destined not to see another U-boat. He was later commissioned and ended the war with the rank of flight lieutenant. Today he still suffers from ear trouble from the shell that burst in front of him.

Seven Days in a Dinghy

It was another 224 Squadron crew that was involved in the next CGM award in September of the same year. To have to ditch in the sea whilst flying was always the great dread of aircrew of any command or nationality. Of course, there were the brave men of the Air Sea Rescue services, the Royal Navy or the lifeboat service always ready to help and answer the call for help.

On many occasions they came out in weather not fit for man or

(*Right*) Sergeant J.S. Powell
(now commissioned)

(*Below*) 224 Squadron:
Sergeant Powell (now
commissioned) 2nd from left
(sitting on chair) Prince
Bernhard in centre

beast but always with the thought that the man or men somewhere out at sea in a dinghy or life raft, were worse off than they were and that their only chance of survival was their coming out to try and effect a rescue.

One man who had to ditch and be rescued and who was later awarded the CGM was Flight Sergeant Ronald John Foss flying as second pilot with 224 Squadron in Liberator FL959 'P'. It was his twenty-fifth operation, and they were on an anti-submarine patrol in the Bay of Biscay. Although based in Cornwall, they had been detached to Gibraltar.

They took off at 10.30 am on 2nd September. At 4.30 that afternoon they were attacked by a long range Ju88 fighter of KG/40, the unit that often sought out Coastal Command aircraft over the Bay. In the ensuing combat the pilot, Flying Officer Wharram, a Canadian, was killed as was another member of the crew, Sergeant Maloney. The aircraft also suffered a good deal of damage. The fighter however, hit by the Liberator's gunners, was claimed as probably destroyed.

After Wharram had been hit, Foss took over the controls and fought to save the aircraft but when, with one engine already useless, another on fire and the two others missing badly, he was forced to ditch into the sea.

Apart from the two men killed, all the crew scrambled into the dinghy. Foss helped to get other wounded men into it and then tended them as best he could. For seven days they drifted, open to the elements. On one day a U-boat was seen and they hailed it but when the Germans discovered they were British airmen, they left them to their fate.

On the 6th day they sighted and were seen by a Sunderland Flying Boat but it was not until the 8th that another Sunderland re-located them. By this time Flying Officer Miller had died in the dinghy. The Sunderland this time dropped some supplies but the men in the dinghy were too weak to reach them and haul them into their fragile craft.

The next day a Catalina found them and guided a sloop, HMS *Wildgoose*, of the 2nd Escort Group to pick them up. While on the way back to England, Pilot Officer Collins and Sergeant Bareham also succumbed to their recent ordeal. The survivors were put

Flight Sergeant R.J. Foss (now commissioned)

Torpedo-carrying Beaufighter

ashore at Liverpool on 21st September. On 2nd October, Foss was recommended for the CGM and Pilot Officer Johnstone, a technical observer in the aircraft, was recommended for the DSO. He had helped Foss to fly the crippled Liberator after Wharram had been killed and assisted with the wounded after ditching.

*

In 1944, Coastal Command flyers received two CGM decorations. The first went to a member of a Strike Wing squadron, flying torpedo-carrying Beaufighters. He was Flight Sergeant Marcus Louis Langley, aged twenty-three from Dunedin, New Zealand, who had been posted to 489 Squadron, RNZAF, as a pilot, on 18th November 1943, having joined the New Zealand air force in 1942.

During an attack on a convoy on 14th May 1944 led by Squadron Leader J.A. Reynolds DFC, 489 Squadron had to run the gauntlet of heavy AA fire. The convoy comprised four merchant ships and no less than sixteen escort vessels, sighted off Ameland in the Frisian Islands. Six Torbeaus of 489 with another six as anti-flak escort, plus twelve Beaus of 455 RAAF Squadron, also flying as anti-flak, headed in.

The Torbeaus had to fly through a veritable wall of AA fire but as they broke away, several torpedoes were seen to hit their targets. One 2,000 ton MV (the Dutch Vessel *Vesta* of 1,854 tons) was hit by torpedoes dropped by Flight Lieutenant T.H. Davidson and Flight Sergeant Langley. There was a huge explosion, then smoke and flame followed. Very soon the *Vesta* was blazing furiously. A second ship was also damaged and an M-class German minesweeper hit by the anti-flak aircraft later sank. One New Zealand Beaufighter failed to return.

Five days later, on the 19th, Langley was flying his fourteenth operation, an anti-E-boat patrol off the Dutch coast. The patrol line was from Den Helder, south past the Ijmuiden, to The Hague. He took off in Beaufighter NE224 'M', with Flight Sergeant Parish as his navigator.

The leader of the patrol was again Squadron Leader Reynolds. They sighted two small vessels of the E-boat type and Reynolds gave the order to attack. No sooner had this been given than they met intense concentration of very accurate heavy and light flak. They were then thought to be R-boats, escorting a convoy which had come out of Den Helder.

In the attack Langley was hit in the throat, arms and thigh, but in spite of these wounds, and seeing two aircraft shot down – Pilot Officer J.A.S. Wright and Flying Officer W.I. Cameron – Langley attacked the target ships before even considering a return to base. Helped by Parish, Langley flew the Beaufighter back to base where he made a successful night landing. Weak from loss of blood, Langley collapsed at the controls as the Beau came to a halt. In the Strike, one R-boat was thought to have been damaged by the intensity of the attack. It seemed likely the convoy had a larger than normal escort, probably being additionally escorted by a heavily armed flak ship.

Langley was recommended for the CGM on 21st May, the recommendation mentioning the operation on the 14th.

It was another Strike Wing pilot that won the next Coastal CGM in August. Warrant Officer Harold Arthur Corbin was born in 1923 and joined the RAF in November 1940, four days after his seventeenth birthday. After training as a pilot he was posted to 235 Squadron flying Beaufighters on anti-shipping sorties from Portreath, Cornwall, but after a few operations he was posted to 248 Squadron at Predannack. One difference, however, was that 248 flew Mosquitos.

His new squadron's main role was attacking ships and aircraft in the Bay of Biscay, the French ports and then later in operations off the Norwegian coast. On 14th August 1944, Corbin took part in a shipping strike on the Gironde River, on the French coast, which led to the port of Bordeaux. He was flying in Mosquito HP866.

Despite heavy AA fire from both ships and land batteries he attacked and damaged a Seetier Class destroyer. However, his aircraft was hit in both outer fuel tanks by heavy flak, the port inner tank was also pierced and all the fuel lost. One shell smashed into the Mosquito through the floor of the fuselage and wrecked the IFF and Gee apparatus.

Corbin set course for Vannes airfield in Brittany, now occupied by the Allies, with fuel streaming from his punctured tanks, the port engine U/S and the starboard engine damaged. When he arrived over Vannes, he climbed to 4,000 feet and ordered his navigator,

Flight Sergeant Webb, to bale out; then he went out himself. Both made successful landings and spent the night under a hedge before making contact with American troops the next day.

Initially he was recommended for the DFC and operations mentioned were an attack on an M-Class minesweeper on 30th June, and an attack on a convoy off the French coast on 27th July, when he was hit and had to return on one engine and a punctured tyre. The recommendation, however, was changed to the CGM by Air Vice Marshal Brian Baker, AOC 19 Group. Flight Sergeant Webb received the DFM.

On 18th September, still flying with 248 Squadron, Corbin attacked *U-867* off the Norwegian coast with cannon and machine gun fire and then dropped depth charges astern of the submarine. The U-boat was sunk the next day by a Liberator of 224 Squadron.

Corbin had been awarded his pilot's wings at the age of eighteen years, five months, and at the war's end he was twenty-one years, six months and a veteran. His award of the CGM came when he was twenty years and nine months. He remembers that the proudest moment of his life was when he went to Buckingham Palace to receive his CGM from King George VI. The saddest moment was when he heard that his old CO, Max Geudj, had been shot down by an FW190 off the coast of Norway. He left the RAF in May 1946 and went into teaching in Australia, but today he has returned to his native England.

*

Warrant Officer Hugh Scott's CGM was not gazetted until October 1945 but by that time he had been on active duty since 1939. At the beginning of the war he was an air gunner with 269 Squadron based at Wick, flying Avro Ansons with Coastal Command. He flew convoy escorts and anti-submarine patrols, and in 1940 flew ops off Norway, around Stavanger, Bergen and Trondheim. The squadron was now flying Hudsons, and at Trondheim, the squadron attacked the battlecruiser *Scharnhorst*.

The squadron continued to operate in Norwegian waters for many months until it moved to Iceland to begin patrols out over the Atlantic. Scott then moved to 502 Squadron, flying Whitleys on similar operations over the Atlantic and also to the Bay of Biscay.

These ops were flown from Limavady, and by 1942 he had some 850 flying hours with Coastal Command, before becoming an instructor.

At the beginning of 1945 he served with 223 Squadron which operated from Oulton, Norfolk, on Liberators and Fortresses. 223 was within 100 Group, flying bomber support missions with radio counter measures. Scott flew twenty ops of this nature including support to attacks on Berlin (twice), Munich (twice), Bremen, Hamburg, the Ruhr, Mannheim and Münster. By April his total flying hours had reached 977. He was recommended for the CGM on 15th May 1945.

CHAPTER SIX

The Night Raiders

August 1943 for Bomber Command proved to be the most active for them during the year. There were a number of raids against Italy in order to help hasten the hoped for capitulation of the Italians from the war, which did come in September.

There was also the important raid on the rocket experimental base at Peenemünde on the Baltic coast. Then towards the end of August, Sir Arthur Harris began what he hoped was the commencement of his next major offensive – the Battle of Berlin. In the event, this battle was to take place later in the year.

It was one of the Italian raids that resulted in an award of the Victoria Cross to one of the participating pilots, Sergeant Arthur Aaron. His bomb aimer was Flight Sergeant Alan William Jessup Larden, who won the CGM.

For Aaron and his 218 Squadron crew, the target on the night of 12/13th August was Turin. Alan Larden was a Canadian, aged twenty-five, and had enlisted in the RCAF in 1941. On the night of the 12th, they took off in their Stirling bomber (EF452 HA-O) at 9.35 pm. It was a long haul to Turin for RAF bombers. For Larden it was his nineteenth raid, Aaron's twenty-first – but their first to Italy. As they neared the target, Larden was in his forward bomb aimer's position and the bomb doors were open. Soon they were on their bomb run flying at 15,000 feet. Suddenly another Stirling ahead of them, no more than 250 feet, began firing. Its rear gunner slowly traversed his rear turret and raked them from wing tip to wing tip. The windscreen was shattered on Aaron's side of the cockpit, the instrument panel and in particular the RPM indicators were shattered, and both front and rear turrets put out of action as

74

three of the four engines were hit, and one put completely out of action.

The navigator, Sergeant Bill Brennan, another Canadian, was killed by a single bullet through his heart and was found lying on the floor amongst his pencils, rulers and maps with his eyes wide open. Sergeant Larden escaped with grazes to his chest and face, lucky to escape serious injury as he was right in the nose of the Stirling, hunched over the bomb sight.

Then Sergeant Malcolm Mitchem, the flight engineer, saw that Aaron too had been hit. He was in a dreadful state. He had been hit in the face, breaking his jaw and a bullet had shot away part of his face. His right arm was only held in place by a few sinews, and he had also been hit in the chest and lungs. He had slumped over the control column but signalled with his good hand for Mitchem to take over. Mitchem grabbed the stick and pulled it back as hard as he could as the bomber had gone into a dive to port and the speed was up to 250 mph. At 3,000 feet, he finally managed to pull it out and got it onto a level heading.

Alan Larden went back and moved the body of the navigator to make room for Aaron. Mitchem, in the meantime, in the right-hand seat, was still trying to fly the aircraft; it was not going to be easy. The throttle control pedestal had been wrecked, the two inboard engine throttle controls had been shot away, and the starboard outer engine was U/S. Losing power all the time, the only engine he could control was the port outer.

While he struggled with the controls, Larden and Sergeant T. 'Jimmie' Guy, the WOP, had carried the wounded Aaron out of his seat and back into the fuselage where they made him as comfortable as they could on the floor, lying against the flight engineer's panel with a parachute pack under his head. His arm was put in a sling and his face wounds covered with a shell dressing.

Larden then went forward again and took over the controls from Mitchem. At first he sat in the left hand seat but with the wind coming in through the smashed windscreen he changed with Mitchem in the right hand seat. The controls were so heavy that he and Mitchem took it in turns to fly the aircraft.

It was clear they needed more height. The tops of some of the nearby mountains were going past them. Larden went forward and jettisoned the bomb load but was unaware that one 4,000 lb bomb had failed to release. They now headed for the Alps but they saw that they were not going to get over the top of the mountains so decided to forget England and turned south for Sicily, but their maps only went as far as the boot of Italy. Then Guy, working on his radio, ·found the frequency and call-sign from the radio station at Bône in North Africa.

Their fuel situation was beginning to cause concern. Larden took over again while Mitchem went to check the fuel gauges, but when he checked on Aaron again, decided he needed to give him a shot of morphine. Sergeant Guy picked up another signal from Bône, advising them not to make for Sicily because of the dangers of high mountains and fighting still going on for the island. Instead they should try to make for Bône itself. It took several hours for them to fly southwards and out across the Mediterranean. Aaron, meanwhile, had come round and was asking the crew how they were doing and how they were navigating. Unaware that a bomb was still aboard, they decided they had better try for a belly landing. Ahead they picked out two signal searchlights in a 'V', guiding them towards the airfield at Bône. To add to their problems, they then received a message that a crashed Wellington was blocking the runway and that in consequence the runway length was down to just 700 yards.

With this news, Sergeant Aaron insisted in returning to his seat to help Larden bring them in. His face was black with caked blood but any advice and assistance he could give was obviously going to help. After making two attempts at an approach, the petrol situation was desperate. They tried again, then again. On the fifth approach Aaron again wanted to abort and try again but Mitchem told him the tanks were dry and so Larden took over and decided it was now or never. They got down, screeched to a halt and luckily the bomb did not go off.

They quickly got Aaron out through the roof hatch, and then went back for the body of Sergeant Brennan. Aaron was taken to hospital but he died of his injuries nine hours later; it had been his chest wounds that caused his end. Who had shot up their Stirling? An enquiry was never held and perhaps that was as it should be. In the heat of battle anything was liable to happen and what was done, was done.

(*Right*) Sergeant Arthur Aaron VC

(*Below*) Aaron's crew; Sergeant Larden in centre

Flight Sergeant Larden was recommended for the CGM on 27th August, and Sergeants Guy and Mitchem received DFMs. Arthur Aaron who had fought to the end was awarded the Victoria Cross posthumously. He was twenty-one and had already been awarded the DFM in October 1943. He is now buried in Bône Cemetery and his VC decoration can be viewed in Leeds City Museum.

Afterwards, Alan Larden was given the safety buckle from his Sutton parachute harness. It had been struck by two bullets, jamming the mechanism in the 'released' position. Had he baled out he would have fallen straight out of it!

Peenemünde

On the now famous Peenemünde Raid on the night of 17/18th August 1943, Sergeant George William Oliver was an air gunner with 467 RAAF Squadron from RAF Bottesford. His Lancaster was engaged by an enemy fighter, on this occasion an Me109 flying a 'Wild Boar' sortie. After successfully bombing the target from 6,000 feet, and leaving the target area, the Messerschmitt came in and attacked. Its first burst wounded the rear gunner, Sergeant Patrick Barry, from Southern Ireland, destroyed the hydraulics and put his turret out of action. Then the Lancaster went into a dive.

At the same time a fire broke out inside the aircraft setting alight the ammunition. Sergeant Oliver, on his twenty-third operation, engaged the fighter from the outset of the attack and continued to do so until the fighter had been destroyed, despite the heat and smoke from the fires below him. By this time it was so intense that Oliver only wanted to get out of what he thought was a doomed aircraft. He clambered to the front of the Lancaster, gestured to the wireless operator that they were on fire and together with the navigator they moved towards the cockpit. Informing the pilot, Warrant Officer W.L. 'Pluto' Wilson, RAAF, of the fire, Wilson said the aircraft was still responding so ordered them back to try and put the fire out.

Oliver went back and with the other two men, worked in relays because of the smoke and finally succeeded in quelling the flames. Then they went back and chopped out the wounded rear gunner. It was due to Oliver's prompt coolness under fire that the lives of the crew and much valuable equipment was saved. For his actions he was recommended for the CGM on the 19th. Pluto Wilson, from New South Wales, received an immediate DFC.

After two more trips – both to Berlin – Wilson and his crew finished their tour and other members of the crew were decorated.

Flight Sergeant Daniel Rees, a pilot with 460 RAAF Squadron, was also detailed to fly the Peenemünde Raid. At the time he was just twenty-one years old and had flown seventeen ops. This was to be his eighteenth and the one he would always remember. Rees hailed from Spearwood, Western Australia, where he was born on 5th February 1922.

He took off at 9.31 pm in Lancaster ED985 and successfully attacked the target at 12.32 am, from 8,500 feet. Over the target his aircraft was attacked by a Ju88 night fighter and severe damage was done to the starboard tail-plane, and the elevator trimming tabs were shot away. The hydraulic system and the starboard outer engine was damaged, also his number one petrol tank in the starboard wing was holed and the petrol drained out.

His gunners, Sergeant 'Chook' Harris and Flight Sergeant John Venning, returned the fire successfully and the fighter was claimed as a probable.

Despite the damage, Rees managed to make it back to base and carry out a safe landing. They had lost 580 gallons of fuel from the damaged No 1 tank. Only five nights earlier, Danny Rees had brought back an ailing aircraft from Milan. He had found his machine suffering from problems on the journey out but persisted in carrying on to the target. He had crossed the Alps by flying through the St Bernard Pass with Mont Blanc towering over his port side. Having attacked Milan he returned in the same manner, despite the engines still giving cause for concern. While crossing the French coast the outer starboard engine had to be completely feathered and the starboard inner carefully handled to keep it going.

Rees received the CGM and was commissioned. He ended the war as a flight lieutenant, having served later with No 7 Communication Unit.

The Big City

Berlin was the target for the night of 23/24th August 1943, when Harris sent a force of 727 Lancasters, Halifaxes, Stirlings and

Stirling Mk III

Sergeant J.C. Bailey CGM on left

Mosquitos to Hitler's capital. In one bomber was Sergeant John Calder Bailey RCAF, a bomb aimer with 622 Squadron. He was born in Canada in 1920, and enlisted in the RCAF in 1941. 622 Squadron had only been formed at RAF Mildenhall on 10th August, so when on the night of the 23/24th, Bailey and his crew took off in Stirling BK816 GI-B, it was only their eighth operation. His pilot was Flight Sergeant Gil Marsh.

They took off at 8.35 pm and bombed the target from 14,000 feet. The route had been a direct course to thirty miles south of Berlin, then a turn to the north for the run in and bomb. They would then return via the Baltic, south across Denmark and home across the North Sea. It sounded easy but there were a few problems which made it anything but easy, and a reception committee that did not want anyone to attempt the journey home.

As Marsh came to the turning point, down went the Pathfinders' markers. Immediately hundreds of searchlights came on, forming a circle of ten miles or so, and, as it was described by Marsh, 'like a huge circus cage. The whole sky over Berlin was as light as day and fighters were seen roaming round waiting for their prey.'

They bombed at 12.07 am, and just after the 'Bombs Gone' call came over the intercom from Sergeant Bailey, a Ju88 came in and opened fire from 500 yards. Both gunners, Flight Lieutenant Berry and Sergeant Hynham, opened fire as Gil Marsh took evasive action. The fighter was hit and claimed as a probable.

Berry was 622 Squadron's Gunnery Leader and had only just arrived on the unit. Sergeant Smith, their usual mid-upper, had been stood down by the MO, so Berry volunteered to replace him.

Another night fighter attacked. Gil Marsh remembers a big bang. His hands were covered in green luminous phosphorus and a 14 pound sledgehammer blow 'hit me with full force. The sciatic nerve was severed at the top of my right leg and there was a large hole at the top of the leg and another through my shattered right hip.' He lost consciousness. When he came to he found the aircraft in a dive and it plummeted from 12,000 to 1,500 feet. Marsh managed to pull it out and climb back to 4,000 feet but no higher as the tail plane and elevators had all been damaged in the attack.

He called to 'Jack' Bailey for help but Bailey had also been hit on the head as he had not been strapped in when they had taken evasive action and so for a while he too had been knocked out. Marsh then told the navigator, Pilot Officer Richards, to get into the second seat and between them they managed to set a course for the Baltic, still flying at 4,000 feet.

The flight engineer called up to inform Gil that the port engine pressure had gone, the engines were overheating and there was an imminent chance of a fire. In the crisis and by mistake, Marsh feathered the port inner engine. When they tried to restart it, it threw out a large sheet of red and orange flame so he shut it down again and they carried on with three engines.

They then got Marsh out of his seat. He had lost a good deal of blood and was pretty well all in. It was then that Jack Bailey, having regained consciousness, took over the pilot's seat. As they reached the coast of Denmark, some flak ships opened up so Marsh ordered Bailey to turn away and fly further north before crossing the Danish coast. When they got out over the sea Bailey and Sergeant Meaburn, the engineer, managed to get the fourth engine going. Then it was a case of flying home.

The problem facing Bailey was the landing. There was no way Marsh could bale out or be pushed out in a parachute with his injuries. He would have to get it down somehow. They reached England and headed for Mildenhall.

Over the airfield they prepared for a crash landing. Marsh was taken to near the rear door, but he had lost more blood by this time and was very ill. To move him would be dangerous. All the escape hatches were opened, ready for a quick exit. Gil Marsh remembers little of the landing as he had gone into shock. He does recall Jack Bailey had been on a pilot course but had failed the exams, and the navigator too had failed a similar course.

As it was, Bailey made a perfect landing, and all were safe. Later inspection showed only 75 gallons of fuel left in the tanks. It had been a close thing. He was recommended for a well deserved CGM on 25th August.

*

Three CGMs were won over Berlin on this August night, the second going to Sergeant Bertram Gordon Bennett who flew as a wireless operator with 623 Squadron. Like 622 Squadron, 623 had only just

(*Above*) Flight Sergeant Gil Marsh and Sergeant Hynham

(*Right*) Sergeant B.G. Bennett GCM

been formed at Downham Market – also flying Short Stirlings.

Bennett was born in December 1913 at Chiswick, Middlesex, and had enlisted in the RAF in 1941. On this night in August, six of the squadron crews were detailed for the Berlin show, one being Flying Officer Overton's, in Stirling EK727 IC-A; his WOP was B.G.Bennett. On this operation, they were joined by Wing Commander Little, squadron CO, acting as second pilot.

They took off at 8.49 pm and reached Berlin without incident to find it illuminated by searchlights and falling fighter flares. At this stage there was little flak as obviously, with the flares in evidence, night fighters were operating in the target area. This was confirmed while on the bomb run. The rear gunner, Sergeant Dallman, saw an Me110 600 yards astern of them and gave the order to corkscrew, which Overton executed. The fighter opened fire from 300 yards using cannon and machine gun. The rear gunner was wounded in the leg while the mid-upper turret was damaged, the sight and firing circuit rendered useless.

The rear gunner had fired at the same moment, followed by a second burst. The fighter ceased firing and fell away in a steep dive with its starboard engine blazing, being claimed as destroyed. During the attack, bundles of 'window' – metal strips dropped to confuse radar coverage on German air and ground radar – in the aircraft caught fire. Sergeant Bennett, who had also been wounded in the back, tackled the fire with his bare hands, until the flight engineer, Sergeant Aubrey, arrived with an extinguisher and together they put the fire out. It was only then Bennett mentioned that he had been wounded.

The Stirling had been holed in the flaps, the mid-upper turret was useless, fuel tanks had been hit and the rear fuselage damaged. Overton flew the bomber back to base and managed a landing despite additional damage to the undercarriage. Bennett and Dallman were taken to Ely Hospital. For his courage, Bennett was recommended for the CGM on the 25th. At the time he was halfway through his tour, but when he left hospital and went on a medical board, he was passed unfit and so remustered to flying control duties, until he left for south-east Asia in June 1945. He left the RAF in February 1946.

The third CGM went to a pilot – Flight Sergeant Osric Hartnell

Sergeant Cowham, Flying Officer White (now commissioned), Flight Sergeant Petrie, Flying Officer Oliver (now commissioned) – all CGMs

White. He had been born in Christchurch, New Zealand, on 21st February 1914 and had been a salesman before joining the RNZAF in October 1941. By August 1943 he was with 75 New Zealand Squadron, and like Bailey and Bennett, was also operating on Stirling bombers on this night. For White, this was to be his twelfth bombing raid. Indeed, he had flown on all three raids during the recent Battle of Hamburg.

On approaching the target area, White's aircraft was coned by searchlights and hit by AA fire many times, but he carried on despite still being coned, then suddenly the flak stopped, which usually meant that night fighters were around. Moments later they were attacked by a Ju88 whose attack killed his rear gunner and shattered the turret. The Stirling went into an uncontrolled dive and White warned the crew to be ready to bale out. In the middle of this the intercom failed, and the navigator, bomb aimer and the wireless operator abandoned the aircraft, simply because they had lost all contact with the captain.

Flight Sergeant White jettisoned the bomb load while still in the dive over the city then managed to regain control when down to

6,000 feet. He then made tracks to get away from the flak and danger area. The flight engineer and mid-upper gunner, the only two left besides the dead rear gunner, and White, took stock of their situation. Despite the damage, the aircraft still responded to the controls but without a navigator they would have problems getting home. Fortunately White had been a keen yachtsman back in New Zealand and knew a bit about navigation plus what he had learnt during pilot training. He decided to fly over Denmark, so headed northwards and from there changed course out across the North Sea. Reaching base he found he had no flap control, landing lights or undercarriage, as the electrical leads had been shot away. They also found they had no radio, so were unable to contact ground control. Despite all these difficulties, White made a perfect landing clear of the runway at his home base of Mepal, Cambridgeshire. His CGM too was recommended on the 25th.

CHAPTER SEVEN

Bomber Command
Autumn 1943

Following major attacks on Nuremburg, Mönchengladbach and Berlin at the end of August, another raid on Berlin and then Mannheim in early September, there was something of a lull in mid-September. However, two raids on Hannover at the end of the month brought more awards of the CGM.

Sergeant Owen Noel Jones was flying with 90 Squadron at Wratting Common as a flight engineer. On the night of 22/23rd September he and his crew, under skipper Warrant Officer Denton, were one of the 711 crews detailed to attack the city of Hannover – never an easy target and always heavily defended.

They took off in Stirling EH944 WP-A, at 6.54 pm and at 9.42 successfully bombed the target area from 14,000 feet. While still in the vicinity of Hannover they came under attack from a Ju88 night fighter whose first burst killed the rear gunner, Sergeant Morgan. The intercom was put out of action and two fires were started at the back of the aircraft. Two of the crew, the wireless operator and the bomb aimer, baled out.

Sergeant Jones went back to fight the fires, despite the fact that the 88 was still attacking, and while doing so was wounded by a cannon shell fragment in his right leg. The pilot too was wounded in the leg and the navigator, Sergeant Suddens, suffered a hand wound. When the fires were out, Jones set about getting the gunner out of the turret, far from easy with his own injury and with a large hole burnt through the rear fuselage floor. Finding the gunner past help, he then moved forward and applied a tourniquet to the pilot's leg and also administered first aid to the navigator.

He found his engineer's panel smashed and the port outer engine had to be feathered owing to shortage of petrol. Later the remaining engines began to splutter, but by switching on all the tanks, opening the balance cocks and operating the wobble pump, he kept the engines supplied with petrol until they reached England. Arriving over RAF Lakenheath, Denton found the undercarriage would not lower, so had to make a belly landing which he did successfully. The Stirling, however, was so damaged that it was written off.

Sergeant Jones was put forward for the CGM on the 26th, and Denton and Suddens were awarded the DFC and DFM respectively. Of the two men who baled out, the wireless operator did not survive and is buried in the Hannover War Cemetery, while the bomb aimer was taken prisoner and held in Stalag 6 (Heydekruge).

Two CGMs in One Crew

Five nights later 678 aircraft returned to Hannover and the night's activities brought a unique award – two CGMs to two members of the same crew. They went to Warrant Officer Arthur Joseph Samuel Walker (pilot) and Sergeant Stanley Mayer (flight engineer) of 101 Squadron, flying Lancasters from Ludford Magna, Lincolnshire.

They took off for Hannover on the 27/28th. As they made their run-up to the target they were coned by many searchlights and engaged by heavy flak and at the same time attacked by a night fighter. The port inner engine caught fire and a blaze started aft of the mid-upper gunner's turret. Sergeant Mayer feathered the burning engine and was later able to restart it.

Warrant Officer Walker also managed to blow out the fire aft by diving the Lancaster, while at the same time taking evasive action which shook off the fighter and got them away from the searchlights. Nevertheless, Walker had to warn his crew to be ready to bale out, but then Mayer went back and put out what was left of the aft fire, but was overcome by fumes and had to be pulled forward by the mid-upper gunner. As soon as he recovered he went back again to make certain the fire was out.

By this time Walker had hauled the Lancaster back up to 15,000 feet, still on three engines but he had to jettison the bombs when in the dive. The intercom had been knocked out, the DR compass was smashed and the trimming cables burnt through. The rear gunner too had been overcome by fumes but Mayer and the mid-upper got him out of the turret, despite a lack of oxygen.

The flight back to base was in thick cloud and on arriving back, now on four engines thanks to Mayer, they were diverted to another base and made a safe landing in appalling weather conditions.

At the time, Walker had flown fourteen operations, Mayer fifteen. Both were awarded the CGM, recommended on the 8th October. On 26th November, they took off on an operation to Stuttgart at 5.13 pm in Lancaster DV285 SR-Q. At 19,000 feet, south of Liège, they were attacked by a night fighter, which started a fire forward of the mid-upper turret. The order to bale out had been given, when the aircraft, already in a steep dive to port, began to heel over.

Arthur Walker, who had by now been commissioned as a pilot officer, baled out and became a POW, but Stan Mayer was killed and is buried in Heverlee War Cemetery, Belgium. He is shown as a pilot officer on the War Grave's register but as sergeant in the squadron records. The navigator, Pilot Officer Jack Blanford, landed some 120 yards from the burning aircraft, but then the bomb load went up. It was 7.40 pm. Blanford had damaged his legs in his landing but was given help in nearby Aywaille and remained in the area until he was fit again. He was then given false papers and set onto a reliable escape line. He was moved into Liège and spent ten days there before setting out to cross into Switzerland. His escape was successful and he arrived in Swiss territory on 8th April 1944.

*

Kassel drew Bomber Command's attention on the night of 3/4th October – 547 aircraft setting out from the English bases. On that raid Flight Sergeant Joseph Vincent Russell, a Canadian from Speers, Saskatchewan, flying Stirlings with 15 Squadron, was down to fly. He headed his bomber, EF459 LS-S, into the target area and came under intense anti-aircraft fire on the bomb run. Finally, at 9.27 pm, their deadly cargo went down from 16,000 feet. Immediately they came under attack from a night fighter. The rear gunner, Sergeant Forrest, returned the fighter's fire, hit it and saw it catch fire and dive straight into the ground. However, the fighter had inflicted a good deal of damage on the Stirling, hitting elevators, rudder, petrol tanks – which were holed – while the Gee

Stirling Mk III in flight

set was demolished and the air speed indicator made unserviceable. In addition, the navigator, Flight Sergeant Burns, was seriously wounded, and the mid-upper, Sergeant Rose, and the bomb aimer, Sergeant Bentley, both received slight wounds.

Sergeant Russell found the aircraft to be tail heavy and tending to stall, so had to use his feet as well as his hands to keep the control column pressed forward. In this way he struggled all the way back to England to make a landing at the fighter field at West Malling in Kent. He received the CGM and was commissioned.

On 20th February 1944, aged twenty-four and by now a flight lieutenant and flying Lancasters with which 15 Squadron had re-equipped in December, he flew a sortie to bomb Stuttgart. His aircraft failed to return, crashing at Botenheim, north west of the city. It was Lancaster LM456. He is buried in the Durnbach War Cemetery in Germany.

On the same 3/4th October raid, another Canadian received the CGM. He was Sergeant William Henry Cardy of 427 Squadron RCAF – known as the 'Lion' Squadron. He came from Toronto and was based at RAF Leeming. He had enlisted in the RCAF in 1940 after serving with the Lorne Scots Regiment for ten months. He trained as an engine fitter and spent 2½ years as ground crew in Canada before sailing to England in 1943. He applied for aircrew duties and later remustered as a flight engineer. He joined the crew of Flight Lieutenant G.J. Laird, and on the Kassel raid – his third – they flew Halifax LK637 ZL-W.

On the outward journey, while still over the Zuider Zee, they were attacked by an unidentified enemy fighter. It raked the Halifax from the port side and below, spraying the length of the aircraft. The rear turret was shot to pieces by cannon fire and the fuselage and bomb bay holed. The gunner, Pilot Officer R.G. Findlay, and the wireless operator, Pilot Officer G.T. Rogerson, were both killed outright, and Bill Cardy critically wounded. He received gun shot wounds to the right arm and left eye yet carried on with his duties until he finally fainted through loss of blood.

In the attack, the 2,000 lb bomb came loose and fell through the starboard bomb door while live. Somewhere between the enemy coast and England, the remainder of the bomb load was jettisoned. During the trip back, Cardy regained consciousness at intervals and gave coherent advice to his pilot and on arriving back at their base, Bill Cardy supervised the emergency lowering of the undercarriage which involved the severing of a hydraulic pipe line, thereby allowing the aircraft to land safely. On the 13th he was recommended for the CGM while George Laird received an immediate DFC.

Hannover was again the target on the night of 18/19th October, the hero of this night being Sergeant Walter Humphrey Cowham. Cowham was a Rhodesian from Fort Jameson, northern Rhodesia. He joined 57 Squadron as an air gunner on 7th October 1943, from No 1660 Conversion Unit, and was quickly into the fray of bomber ops. By the 18th he had flown five raids, and on the Battle Order for Hannover that night. His pilot, Flight Sergeant Grimbly, from Perth in Western Australia, lifted off their Lancaster from East Kirkby, Yorkshire (EE197) and headed towards Germany.

Hannover was a hot target, and they could expect a good reception by the German flak gunners and from night fighters – as they were soon to find. Sitting in the rear turret, cut off from the rest

Flight Sergeant W.H. Cowham CGM

of the crew, Walter Cowham kept vigil. It was even stranger for him as this was the first time he had flown with this crew. Gunners would often fly as spare crewmen and fit into other crews whenever needed, if injury or illness depleted another crew. Being a spare gunner on any unit was not the best of situations.

However, they reached Hannover, made the approach and the bomb aimer, Sergeant Firth, put their bombs down on target. It was now time to get away and home as quickly as possible as the fighters would soon appear amongst them along the bomber stream now that the target was known. When only a few minutes off target, Cowham reported a Me109 trying to make an interception. In the words of Flight Sergeant Grimbly:

We had just closed the bomb doors when two fighters came in and attacked us. First the mid-upper, Sergeant Fox, who hailed from Yarmouth, shouted, 'Fighter on the port quarter!', then Sergeant Cowham called, 'I have got him covered, Skipper'.

He then swung his turret to starboard to have a last look in that direction when he shouted, 'Another on the starboard!' He had seen a Me109 coming in to attack at a distance of about 100 feet.

We were a sitting shot for the enemy fighter at that distance and he raked us with cannon and machine gun fire from one end of the aircraft to the other. Though he was hit by the first shell, Sergeant Cowham replied with his guns and saw hits on the fighter. It dived straight down underneath our tail and vanished from sight. As this fighter broke away another came in on the opposite side, and when the mid-upper fired at it this also went into a dive and was lost to view.

Sergeant Cowham had in fact been blinded in his left eye by a splinter from a cannon shell which burst inside his turret. He refused, however, to leave his post. Another shell had gone through the radiator on the starboard outer engine which quickly seized up and caught fire. Yet another cannon shell hit one of the fuel tanks on the port side and 120 gallons of petrol had been lost, while a fourth shell went through the starboard tail plane. Grimbly continues:

We had other hits at various places along the fuselage and the windscreen in the nose was riddled with holes made by machine gun bullets. One of these went ripping past my leg so close I felt its movement. A piece of cannon shell also went through our dinghy, although we did not know it until afterwards. It was fortunate for us we did not have to ditch.

About 25 minutes later a Me110 came in on the starboard side and fired at us. We dived and got away from that one, then again, near the Dutch coast another Me110 came in from astern. Sergeant Cowham fired at this and drove it off before it came near enough to fire and possibly hit it. Over the Dutch coast I asked the engineer, Sergeant English, if we had enough petrol to reach base at East Kirkby. He said no, and so I set course for the nearest aerodrome and landed safely.

When we got Sergeant Cowham out of the turret, apart from his eye wound a bullet had passed through his flying suit and taken some skin off his shoulder. We also found that Sergeant Fox had been hit in the foot. 'Mike', as Cowham was known to his friends, had lost a lot of blood but despite all pleas he would not vacate the turret until we landed.

In Rhodesia he had been keen on big game hunting and when he was seen in hospital by the crew he told them – 'I won't have to squint my eye now when I go lion shooting again!' His CGM was gazetted on 19th November.

Leipzig

A raid against Leipzig on the night of 20/21st October brought two more CGM awards – both to pilots. It was the first serious raid on this distant German city and weather conditions were against the bomber crews. They were later described by the returning airmen as appalling. In consequence the bombing was somewhat scattered and only 271 aircraft of the 358 despatched actually bombed. All were Lancasters of 1, 5, 6 and 8 Groups. Sixteen failed to return.

Warrant Officer Claude Edward White began flying with 100 Squadron in May 1943, going on his first operation – to Dortmund – on the 23/24th. From then until October his targets had been Düsseldorf, Essen, Cologne, Munich, Berlin, Hannover, Nuremberg, Milan and Hamburg during the Battle of Hamburg. On this October night he took off for Leipzig at 5.58 pm in Lancaster DV189 HW-B2.

While climbing over the airfield after becoming airborne, their troubles began. The intercom went u/s and from there on they had to rely on the emergency use of the radio transmitter equipment. Soon after crossing the enemy coast a mechanical fault rendered the port engine u/s and before it could be feathered it caught fire. It continued to burn brightly enough to illuminate the fuselage and tailplane. Despite this and being now an easier prey for fighters, they continued on to the target.

After reaching and bombing the target, they set course for home about seventeen minutes behind the main force bomber stream – and the port engine was still leaving a trail of flame. Yet they managed to get back; on arriving over their home base, people on the ground could see the burning engine. On landing the flames flared up and were threatening to engulf the whole Lancaster. White ordered his crew out while he stayed and dealt with the petrol cocks and switches.

He was recommended for the CGM on the 25th for his bravery in pressing on when he would have been quite entitled to abort the mission. Even when over home base he first let another aircraft land ahead of him as it had been flashing an SOS signal! Even after completing this hazardous operation he and his crew were on the battle order to Kassel the next night – his twenty-ninth mission. He was later commissioned, and while serving as an instructor with No 30 OTU, he was awarded the Air Force Cross, which was gazetted in the New Year Honours List in January 1946. He served as an instructor for seventeen months and undoubtedly any pupil seeing their instructor with the CGM ribbon on his tunic, must have felt in very good hands.

The second pilot decorated following the Leipzig raid was Flight Sergeant Frederick John Stuart, who came from Fourstones in Northumberland, where he had been born in 1916. He flew with 426 (Thunderbird) Squadron RCAF.

On the 20th his name appeared on the battle order and he took off in Lancaster DS686 OW-D, at 5.15 pm. Before they reached the target they were attacked by a night fighter, not on one, but on seven occasions!

It was the first seen at a range of 300 yards as it opened fire on them. The rear gunner, Flight Sergeant Andrew, yelled for a diving turn to starboard as he too opened fire. The fighter, identified now as a single engined Me109, scored hits on the tailplane, fuselage and wings of the Lancaster, rendering the mid-upper turret useless while wounding its gunner, Sergeant McGovern. The fighter made a further six passes, but each time evasive action by Stuart frustrated the enemy pilot and they were not hit again.

Taking stock after the attacks, the damage was found to be quite extensive. Besides the upper turret, the Gee apparatus was useless, Stuart's windscreen had been shattered, the trailing aerial was shot away, holes were all over the wings and fuselage and the controls did not feel 100 per cent. However, they carried on and bombed the target at 9.03 pm then made tracks for home, Stuart making what was described as 'a masterly landing'. His CGM recommendation was put forward by the squadron CO on the 23rd, Stuart having completed twenty operations.

On 20th December Stuart set out in Lancaster LL630 to bomb Frankfurt but was among the forty-one aircraft that failed to return. He is recorded as being buried in the Rheinburg War Cemetery in Germany.

CHAPTER EIGHT

The Battle of Berlin

Over the winter of 1943-44, Sir Arthur 'Bomber' Harris finally managed to wage his war on Germany's capital city – Berlin. Bomber Command sent major forces to Berlin on no less than sixteen occasions between November 1943 and March 1944. Four men would win CGMs during these raids.

Yet it was against Düsseldorf on the night of 3/4th November that brought the first two of this winter's CGMs. Sergeant Thomas Ernest Bisby was a wireless operator with 10 Squadron at Melbourne, Yorkshire. He had joined it in May but had only carried out ten operations by November, flying in Halifax bombers.

On the outward route to Düsseldorf, in a matter of just four minutes, they were attacked by no less than four night fighters. The first one seen was an Me210 coming in from the port quarter. Sergeant Smith in the rear turret, reported it to the pilot, Flight Lieutenant R. Trobe, who corkscrewed the bomber to port. Smith opened fire and the fighter did likewise, damaging the intercom and call light. The Messerschmitt broke away below.

At the same moment another fighter, this one unidentified, came in from above and opened fire at 400 yards, and this time the rear turret was knocked out, and the gunner down to only one gun which he had to fire manually. The mid-upper was still in action however. The 210 then attacked again from above. This time the upper gunner, Sergeant Mowatt, claimed hits on it and flames were seen coming from one of its engines. Another fighter attack came in but this was successfully evaded.

In addition to the intercom and rear turret, the wireless had been hit, the hydraulics knocked out, bomb doors and flare chute damaged, the fuel tanks in the port wing damaged and the port inner engine stopped. The flight engineer, Sergeant Bridge, was also hit and wounded, and Tom Bisby had been badly wounded in the legs and had a tough job on hand to repair his damaged radio. Trobe headed the damaged bomber back towards England. In pain, Bisby repaired his radio and then obtained a wireless fix which turned out to be their only navigational aid, as well as passing a message to Group HQ. The radio then broke down again but he once more repaired it. When they reached Woolfox Lodge, Rutland, they were given permission to land but as the radio had gone u/s again, Bisby had to signal with the Aldis lamp. To do this he had to stand, although he needed the assistance of other crewmen to do so. For his courage and determination, he was put forward for the CGM on the 8th. He went on to obtain a commission in October 1944 and became a flight lieutenant in 1948. It is sad to record that Tom Bisby died of a heart attack in the 1970s.

There was also a Victoria Cross awarded for heroism on this night, which went to Flight Lieutenant William Reid of 61 Squadron. It was his eighth operation and his flight engineer was Sergeant James William Norris.

They took off in Lancaster LM360 QT-O at 4.59 pm from RAF Syerston to fly to Düsseldorf. Only minutes after crossing the Dutch coast at a height of 21,000 feet, they too were intercepted by a night fighter. The rear gunner, Flight Sergeant A.F. 'Joe' Emerson saw the 110 and had it in his sights but his fingers were frozen as his electrically heated suit had failed; as the fighter got to within 150-200 yards he finally managed to fire and observed hits on it. However, the fighter managed to get in a burst which put his turret out of action. The intercom was also hit. Three to four seconds later another attack took place, this time by an FW190 which attacked from the port beam and fired two 3-4 second bursts from 400 yards. Emerson managed to get off a burst but now he had only one gun working, but with the state of his turret, accuracy was almost impossible. The Lancaster's fuselage was raked by cannon fire and the mid-upper turret put out of action.

Bill Reid, already wounded in the first attack, was hit again, and Sergeant Jim Norris wounded in the arm which later turned out to be broken. The navigator, Flight Sergeant J.A. Jefferies, was killed although this was only discovered later when his body slid forward into the cockpit area. The wireless operator, Flight Sergeant J.J. Mann, was mortally wounded.

All the wireless equipment and flying instruments were knocked out, the oxygen system was out of action and both compasses u/s. Norris kept Bill Reid going by passing him portable oxygen bottles.

The aircraft became very heavy on the controls and tended to fly nose down, as the elevators too had been damaged. It required, on Reid's part, a great deal of heavy and continuous pressure on the control column to keep the Lancaster on an even keel.

They were now down to 20,000 feet and, with a smashed windscreen to contend with, it got very cold. In these severe conditions, Norris helped the pilot to keep on course for the target for fifty minutes and he was often forced to take his oxygen mask off owing to the effort needed just to keep the control column back. In doing so, Reid had to face a freezing gale of air coming right at him with a temperature at that height, of 20° below.

After successfully attacking the target, Reid turned for home but needed even more help from Jim Norris. Loss of blood from his wounds was beginning to affect Reid and on more than one occasion, the Lancaster went into steep dives and on one occasion did two circles in a diving turn before control was regained.

They navigated by the moon and the English coast finally came into view. They saw a lit-up runway, which turned out to be Shipham. It was only then that Norris mentioned that he had been wounded. By now he had been helping to hold on so long to the control column with Reid, that his arm had gone stiff. Blood from Reid's head wound threatened to blind him as he circled the field, flashing his landing lights to indicate his distressed condition. He had now to get the wheels down, using an emergency pressure bottle, but the effort only helped to reopen his wounds to head and shoulder. However, the wheels were down which was just as well as the bomb doors were still hanging down from their bomb run.

Les Rolton, the bomb aimer, stood behind Reid, ready to grab Reid if he fainted, as the rest of the crew took up crash positions. As they touched down the undercart collapsed and the bomber screeched along the concrete runway until finally stopping. The airfield was the home of the 44th Bombardment Group, USAAF and they had landed just on 10 pm.

Sergeant J.W. Norris CGM

One damaged Lanc

Bill Reid VC on right

Flight Sergeant Mann died the next day. Reid and Norris had been taken to Norwich hospital from where they learned that they had been awarded the VC and CGM respectively, while Joe Emerson received the DFM.

The Battle of Berlin

The first of the Berlin series of raids took place on 18/19 November. Raid number four occurred on the 26/27th. For Sergeant George William Meadows, a Canadian from Winnipeg, of 166 Squadron, this raid to the Big City was only his third operation.

On this occasion he was flying as rear gunner with Flight Sergeant Fennell in Lancaster DV365. Their take-off time was 5.23 pm and their main bomb load was a 4,000 lb 'cookie' plus a number of incendiaries. It would be a long haul and the petrol needed to get them there and back meant that the bomb load had been reduced.

On the outward journey over France they were attacked by an enemy fighter. It was the worst kind of attack – head on! The mid-upper, Sergeant Cushing, was wounded and his turret put out of action for a time. The German immediately came in for a second run and this time a cannon shell exploded in the front cockpit. The elevators were damaged which in turn caused the aircraft to become uncontrollable. The bombs were jettisoned and the order given to bale out, but then Fennell managed to regain control.

In the rear turret, Sergeant Meadows had also been hit and his turret damaged. A bullet entered his back and having been deflected by the wiring of his heated suit, came out at his left groin. Ignoring the severe pain of his injury, he continued to direct Fennell as the fighter came in again, and he also beat off further attacks, firing his guns on several occasions while doing so. In all they were attacked twelve times and it was amazing that they survived and made it back to make a successful landing at Ford, in Sussex.

When the order to bale out was given, the bomb aimer in fact did just that before Fennell regained control. The aircraft had been badly damaged as one would expect and in addition to the two gunners, the navigator also suffered a slight wound. For his courage, Meadows received the CGM, recommended on 30th November.

Berlin raid number five came on 2/3rd December. Warrant Officer Edward Sydney Ellis, born in Luton in 1914, was flying with

625 Squadron on this night. He took off in Lancaster DV362 from Kelstern in Lincolnshire, and bombed Berlin at 8.22 pm from 20,000 feet. On the bomb run they were hit by flak and the rear gunner, Sergeant Wightman, was wounded. The run was continued and as the bombs tumbled down, a fighter attacked from below, raking the aircraft from stem to stern.

Once again Wightman was hit and the mid-upper, Sergeant Jones, was also wounded. Still on the bomb run, Ellis kept straight, continuing the delay between dropping their high explosive bombs and their incendiaries, which had to be carefully timed. Only after the latter had been dropped did Ellis take evasive action to lose the fighter. The intercom was then found to be u/s and the bomb doors would not close, so the hydraulics too must be damaged. As they were to discover later, the main wheel tyres had been punctured in the fighter attack.

They were also short of oxygen so Ellis reduced height to 12,000 feet but when crossing out over the Dutch coast, the aircraft was hit again by flak, and the engineer reported that they were losing fuel. Reaching England, Ellis selected an airfield about twenty miles from base, as the fuel state was fast becoming critical. Using the emergency pump to get the wheels down, Ellis came in without flaps and with flat tyres. As it touched down, the Lancaster nosed forward but tipped back again and nobody was injured. They had landed at RAF Bardney, the home of 9 Squadron. Both gunners were taken to hospital where they recovered.

Ellis was recommended for the CGM on the 5th and was later commissioned. After thirty-six operations and now a flight lieutenant, Ellis received the DFC. In the recommendation for this second award, mention was made of an attack on Leipzig on 19th February 1944. He lost an engine while still some distance from the target but in true tradition, carried on, bombed and returned home on three engines. During his tour, which spanned the Battle of Berlin, Edward Ellis flew on eight of the sixteen attacks on the city.

Warrant Officer Richard Jack Meek went to Berlin on four occasions. He was a Canadian born in Vancouver, British Columbia, in 1908. He enlisted in the RCAF in 1941 and at the age of thirty-three, trained as a navigator.

On the night of 30/31st January 1944, he was detailed to fly on his

Warrant Officer E.S. Ellis CGM (now commissioned)

sixth operation with his 626 Squadron crew out of RAF Wickenby. The squadron had only been formed in November the previous year, from C Flight of 12 Squadron, also at Wickenby.

Berlin was a long haul through fighter-infested territory and when the crews arrived over the city there was usually a rousing reception from the flak gunners. As too it was winter time, the weather often proved adverse for flying adding new dimensions to the tension and danger of operational bombing missions to Germany.

Meek's Lancaster, LM584 UM-Y2, took off at 5.14 pm with Pilot Officer Breckenridge in the 'front office' seat. They reached Berlin without problems, but on the bomb run they were attacked at close range by a fighter. Its initial burst of fire wounded the rear and upper gunners, Sergeant Schwartz and Pilot Officer Baker, and killed the wireless operator, Sergeant J. Hall. The aircraft was badly damaged. The hydraulic system, oxygen system, elevators and rudders were damaged, the bomb doors would not close, several of the pilot's instruments were shattered or made u/s, one fuel tank was holed and the electrical system burnt out.

After taking evasive action they returned to the bomb run and dropped their 'eggs' from 20,000 feet. Just afterwards they were attacked again by a fighter which fired a five-second burst. With both gunners out of action, the pilot was given no warning but he did his best to evade. Warrant Officer Meek was hit in the chest by a bullet, striking very close to his heart, while another pierced his left shoulder blade. Two further attacks inflicted more damage on the bomber.

Despite pain and loss of blood, Richard Meek managed to muster enough energy to plot a course for the pilot to get home, even though most of his navigational aids had been destroyed or put out of action. He remained at his navigating table throughout the return trip and kept them on track.

While crossing the North Sea the electrical system caught fire which took the last bit of technical navigating aid Meek had, away from him. He only left his position when an RAF base had been located and identified. This was at RAF Docking. Such was his accuracy in navigation that the aircraft was plotted as being only three miles north of its correct track when leaving the enemy coast.

Breckenridge had to make a belly landing but it was successful and no more injuries were sustained. Meek and Schwartz were both taken to the RAF hospital at Ely, and Pilot Officer Baker was admitted to Docking's sick quarters. The rest of the crew after a rest and some medication, returned to Wickenby.

Meek's CGM was forwarded for endorsement on 3rd February and after recovering from his wounds, he returned to operations. On 1st August 1944, now a pilot officer he was recommended for the DFC, having completed twenty-eight operations.

The 15th Berlin raid took place on 15/16th February by a force of 891 aircraft, after a break of more than two weeks on the capital. On this occasion it was another air gunner who won the CGM.

Flight Sergeant Geoffrey Charles Chapman Smith was an Australian, born at Marrickville, Sydney, on 3rd February 1919. As a member of the local militia, in the 18th Field Bde, 56th Battalion Artillery, as a second lead horse driver he was already in uniform when war began. He had joined them in 1938, their guns being left over from World War I, and remained with this unit until 1941. The following October he enlisted in the RAAF, trained in Australia, South Africa and England, then with the rank of sergeant, was posted to the new 625 Squadron in November 1943.

He remained with them until 2nd January 1944 when he was posted to 7 Squadron, but as he arrived he was immediately reposted to 156 Squadron of the Pathfinder Force. While with 625 he had quickly made his mark, shooting down a Ju88 on his first operation over Berlin on 2nd December (the 5th raid of the series) – the same operation on which Ellis received his CGM, while with the same squadron.

On 15th February, he and his crew were on the Berlin battle order once again. They took off in Lancaster ND444 GT-O, with Flight Sergeant Ken Doyle, a Londoner, in the pilot's seat, at 5.26 pm. When thirty miles from Berlin on the outward trip, Doyle received a message on the intercom from Sergeant Smith. He reported a fighter coming in behind and gave evasive instructions. Smith then began firing and the next thing Doyle saw was tracer shells flashing past over his head. The crew then saw the Me110 flashing its navigation lights on and off, and also waggling its wings. They thought it must be a decoy so the upper gunner made a search for another fighter. Sure enough another fighter was around but it attacked without them seeing it. The aircraft was hit between the upper and rear turrets, wounding both gunners, Sergeant Clarke in the upper turret having a compound fracture of the left leg caused by a cannon shell, while Smith was hit in the right ankle by a shell which exploded and severely wounded him. Both turrets were out of action because of the hydraulic damage.

Smith reported to the pilot that the 110 had been hit by their fire and had exploded in the air so was being claimed as destroyed. Then another fighter, this time a FW190, was spotted by Smith. This attacked, but its fire went below the Lancaster and missed.

All the crew were called on the intercom but there was no reply from the mid-upper. Sergeant Clarke, as well as being wounded, had had his intercom knocked out in the attack. They were still heading for Berlin but then found they could not open the bomb doors and so reluctantly, when only fifteen miles from the city, they decided to turn back.

The navigator, Sergeant Winlow, from County Durham, then called to the pilot that a fire was burning in the back of the aircraft. He went back with an extinguisher. It turned out to be Smithy's parachute that was burning in its stowage position. By this time, Doyle could hear Smithy groaning over the intercom. Sergeant Alf Astle went back with an axe to get him out of his turret, but Smith would not hear of it and said that as the mid-upper was u/s he needed to stay at his post and keep a look out for fighters. Regardless of the pain, he was still able to manipulate his turret by hand.

Their troubles were still not over, as they ran into flak, and a lump of shrapnel hit the throttle box by Doyle's feet and severed the controls to the port engines. In the meantime, Sergeant Syd Richardson, the flight engineer, managed to reach the forward hydraulic jacks on the bomb doors via the bomb bay inspection hatch. He dismantled the connections on the hydraulic system in order to release the pressure. This would allow the doors to open when they jettisoned the bombs on top of them. Over the sea they tried it, and all fell through and away except one 500 lb bomb which refused to go, despite several manipulations of the release and jettison handle, so they brought it back.

(*Left*) Flight Sergeant G.C.C. Smith CGM (now commissioned) (*Right*) Geoff Smith, Australia 1983

Flight Sergeant Smith on crutches at Buckingham Palace

Once well out over the sea, Astle and the wireless operator, Sergeant Don Green, went back and chopped down the doors to Smithy's turret. It took them half an hour. They then got him out and back to a safe position for a crash landing. Doyle was heading for Woodbridge on the Suffolk coast and was given permission to land. Green, who had gone into the upper turret when Clarke had been taken out, got down and with Winlow and Astle lay down and wedged themselves against the two wounded gunners to keep them in position.

The landing, thanks to Doyle, was very successful at Woodbridge, at 12.55 pm. Besides the wounded gunners, Sergeant Green suffered from frostbite, having lost his gloves while tending to Geoff Smith. Smithy was taken to Ely hospital where his right leg was amputated above the knee. For gallantry in staying at his post to protect his aircraft and crew, he was recommended for the CGM on the 18th.

In May 1945 he returned to Australia where he worked with the Repatriation Department as an artisan and surgical fitter. Later he became a sales manager for a multi-national organisation and stayed with them for twenty-five years. He played lawn bowls and snooker, also, occasionally, golf. He also danced but in his own words admitted to 'no fancy steps'.

Only two days after losing his leg he realised the problems that lay ahead and said to himself – I am as good as the next guy until they prove themselves better than I. This doctrine and the support of his family, plus his own independence in doing things for himself, saw him through.

The story of Geoff Smith ends on a sad note. Three of his crew, Ken Doyle, Alf Astle and Donald Green, were all lost in September 1944 on an operation over Calais. It was their last operation and they were seen to be hit by flak and crash into the sea. Their bodies were never recovered but they are remembered on Runnymede Memorial for the missing. Geoff himself died on 6th February 1986, having been ill for some time previously.

*

Five nights after Smithy's last op, Bomber Command returned to Leipzig, but it cost them dear. No less than seventy-eight heavy bombers failed to return out of the 823 despatched. Night fighters were very active and a diversion raid to Kiel Bay failed to draw them sufficiently from the main force. In this way the bomber stream was under attack all the way to the target. The forecast winds were also out, causing navigational difficulties for the crews.

Operating this night was Sergeant Barry Colin Wright, who was a flight engineer with 166 Squadron, in Pilot Officer Catlin's crew. They took off in Lancaster LM382 AS-Q, at 11.40 pm and soon after crossing the enemy coast, Catlin and his navigator soon realised they had a strong tail wind, which meant they would arrive over the target too early. Catlin started to dog-leg to lose some of the time, although they knew there would be many of the 800-odd aircraft about them doing much the same thing.

Nearing the target they were attacked by a Me110 fighter. The rear gunner, Sergeant Bill Birch, began firing before he had finished his alert call to the pilot. In spite of his efforts, they were raked from rear to front with cannon fire. The mid-upper, Sergeant Tom Powers, navigator, Pilot Officer Tony Pragnell, and Wright were all wounded.

A cannon shell burst in the area of the rear turret on the port side, scattering .303 ammunition from the bulkhead to the turret. It also blew a hole in the port side of the fuselage about three and a half feet square. Most of the perspex in the upper turret was blown out. Another shell ripped through the large electrical panel on the starboard side of the aircraft and fused every light from front to back.

No sooner had this shock attack hit them, than another 110 came in, so the bombs were jettisoned. Billy Birch fired on the second Messerschmitt and it was seen to go down in flames. The first 110 re-appeared but was soon sent packing by dead-eye Birch in the rear. Sergeant Wright, meanwhile, had collapsed into unconsciousness, but when he came round his first thought was to get on with his job. His control panel had been shot away and one petrol tank had been holed and drained of its contents, but he succeeded in keeping all the engines running at maximum power. Catlin tried to persuade him to lie down and rest but he refused. On no less than three occasions he fainted through loss of blood.

The bomb doors were still open as the hydraulics had gone, and the trim controls were locked and set for a full bomb load so the aircraft continually wanted to climb. Catlin having to fight it all the way home.

On reaching England they found their own base was closed due to bad weather so they were diverted to the emergency field at Manston. On arrival the undercarriage was locked down manually and, if this was not enough, they then saw that both tyres were flat. Catlin made a three-point landing more or less on the rims, in a shower of sparks, but he got them down safely. After landing the mid-upper gunner had to be cut out of the turret with an axe. His flying suit too had to be cut off him and fragments of metal clinked to the fuselage floor as they did so. He was then taken to hospital.

Barry Wright too was helped on to a stretcher. He could not stand up straight and could only move with assistance. The navigator had shrapnel in his shoulder and near to his neck. The wireless operator had a piece of shrapnel lodged in the back of his upper arm, but both men were well enough to return to the squadron the next day.

The aircraft was so badly damaged that the salvage unit broke it up where it stood.

All seven men were decorated. Barry Wright received the CGM. Jim Catlin, Pilot Officer Fred Sim, the bomb aimer, and Tony Pragnell each received the DFC. DFMs went to Tommy Hall, the wireless operator, Tom Powers, the mid-upper, and Sergeant Billy Birch.

*

Not all bomber squadrons flew night raids into Germany and the next story concerns one unit which flew what became known as 'clandestine' operations.

Sergeant Herbert Allison Donaldson was born in Selby, Yorkshire, in 1922. In November 1940 when he was eighteen, he volunteered for the RAF but was not called up until January 1941. The next year he began wireless operator training which he completed satisfactorily. He then started aircrew training, first on Botha aircraft, then Wellington, and then Stirlings.

In October 1943 he joined 199 Squadron at Lakenheath. Here he flew on a variety of operations, mining, bombing, etc., but the operations he liked the most were dropping supplies to the Maquis forces in France. It was very satisfying as one could see the results of

Sergeant H.A. Donaldson CGM (now commissioned)

their efforts. Men could often be seen running on the ground, waving and flashing torches at them, as they recovered the arms and supplies parachuted down. Often the next day, the CO would confide that, 'They got the stuff all right last night, chaps.'

It was on one of these operations, on 5th March 1944, flying in Stirling EJ115 EX-C, flown by an Australian, Flight Lieutenant Barson, that things got a trifle hot under the collar.

They had just passed over the enemy coast when they were hit with bullets and shrapnel, probably light flak or even machine-guns as they were flying in low. The wireless was set on fire and a bullet went through his desk and ignited a number of Very cartridge flares which they carried in order to try and fool the enemy with false colours of the day. Donaldson himself collected shell splinters in his shin and thigh, but he grabbed an extinguisher and with the help of the engineer, Sergeant Wild, put the fire out. The aircraft was full of smoke and when he returned to his Marconi set, found it totally dead. He tried every check in the book and switched valves over but it would not work and he could not find out why, yet when he reassembled it, it suddenly worked. Having told Barson, they decided to press on and complete the job. It was then that he mentioned to Barson that he had a hole in the leg, but not to worry.

In great pain he kept at his job for some hours until they landed back at base at 2.59 in the morning. He was taken to Ely hospital by ambulance, but his stay there, as he recalls, was a short one. He was recommended for the CGM on 8th March. He returned to flying duties with 199, with whom he had now flown eleven operations, and continued with them until December. At this time 199 were flying from North Creake, having moved there in June. Bert Donaldson remembers:

> Every time we landed away from base we would be asked where we came from. After a couple of months we had heard every possible variation about a certain 'creek' and paddle!

When he left the squadron he flew in Anson aircraft at an Advanced Flying Unit, remembering it as being very 'hairy' flying around Scotland. As Ansons could not maintain height on one engine and with mountains, etc, all around, it was not the best of places to fly. He left the airforce in May 1946 and now lives in Yorkshire in happy retirement with his wife, having three children and two grandchildren.

CHAPTER NINE

Middle East Heroes

A number of men received Conspicuous Gallantry Medals for bravery during operations in the Middle East and in Italy during the period 1943-1945. The first two such awards went to members of the same crew flying bomber ops in North Africa.

Sergeant John Patrick McGarry was born in Romsey, Hampshire, in 1921. Prior to his RAF service he had been a window dresser, a rather remote occupation from that of a navigator on 70 Squadron, flying Wellington bombers in the desert.

Before his CGM operation he had quite a scare after taking off from Gardabia West, in Libya, en route for Mahares. They had to return when their port engine cut just after take-off, always the worst possible time for this to happen. The bombs were jettisoned within two minutes of becoming airborne and they crashlanded at a speed of 70 mph. The aircraft was completely burnt out but the crew had only slight injuries, which shows the skill of their pilot, Sergeant Petrie.

Thomas Parker Petrie, aged twenty-nine, was from Dundee, Scotland. Just one week after this hair-raising experience, Petrie, with the same crew, took off in Wellington HF753 DU-O, for Menzel Temime, an enemy landing ground. The date was the night, 12/13th April 1943.

On the run-up to the target they were hit by AA fire. Despite this, Petrie kept on course and released a stick of bombs across the target. Immediately they were hit again and the Wellington went into a steep dive. Petrie had been badly wounded and his left foot had been almost severed, while Sergeant McGarry, too, had been wounded in the right leg by shrapnel. Petrie got the aircraft back under control, but then the bomb aimer, Sergeant Bennett, helped by Sergeant Smith, the rear gunner, and McGarry, got Petrie out of his seat and gave him first aid. Bennett took over the controls while McGarry, still trying to make Petrie as comfortable as possible, continued to navigate, giving the bomb aimer the correct course to steer in order to avoid flying over enemy-held Tripoli.

When they got near their own base, Petrie gave the crew the chance to bale out but they decided to carry on and attempt a forced landing. Sergeant Petrie was helped back into his seat. To do this, part of the cabin was cut away and once in position, his right, uninjured leg, was strapped to the rudder bar. As he came into land at a height of 300 feet, their petrol ran out, yet he made a successful crash landing.

By this time McGarry was suffering from exhaustion and loss of blood plus shock which set in quickly once the immediate danger was over. All escaped further injury, but the next day the gallant Petrie had his left leg amputated. Both men were recommended for the CGM on 19th April.

Sergeant McGarry went on to fly again but on 2nd July, his aircraft was reported missing after an operation to Kairouan Temmari in Tunisia. A Wellington was seen shot down $3\frac{1}{2}$ miles south of Olbia, which would have been near the target area of the Olbia marshalling yards. He has no known grave, but his name is remembered on the Malta Memorial, Panel 9, Column 1.

*

Just a week after Petrie and McGarry's fateful trip, on 21st April, an award went to an Australian air gunner of a reconnaissance unit. In January 1941, George Allan Downton, from Adelaide, enlisted in the RAAF when he was twenty-three years old. He undertook his initial training in Australia and then Kenya and Rhodesia before finally arriving at an operational unit, No 1437 Strategical Reconnaissance Flight, in the Middle East, in July 1942. Just four days before his final war flight he was promoted to the rank of warrant officer.

On 21st April 1943, he and his unit were based at El Djem and he was detailed as air gunner to fly on a reconnaissance to the Cape Bon area. His aircraft, a Baltimore (AG766), took off at 4.15 pm, with Lieutenant Ballard, a South African (SAAF), in control.

Besides the normal crew of four, they were carrying a passenger, Major Braithwaite, another South African, attached from No 40 SAAF Spitfire Squadron. When dusk arrived they were overdue. A flarepath was lit in readiness for a late arrival and the air sea rescue services were alerted, but to no avail; they were posted as missing.

The next day German radio messages were intercepted by 211 Group, that a dinghy had been picked up. It was hoped it was Ballard and his crew but for a while nothing further was heard. The story of what actually happened, however, remained a mystery until July, when Downton was reported a prisoner.

They had in fact been attacked by ten Me109s. In the air battle Lieutenant Ballard and the other gunner, Flying Officer Tassie, were killed, and the observer Flight Sergeant Hartley, badly burned, so badly that he died the next day. George Downton engaged the attacking fighters but found his gun was not working properly so directed his energies in directing Ballard during the running fight, which lasted twenty minutes. Downton was slightly wounded as the aircraft was forced down. When Ballard was hit, they were over the Bon peninsula, and the Baltimore crash landed.

Downton scrambled out, and regardless of the fire and flames from the burning aircraft, he went into the wreck and helped extricate the badly injured Hartley. He then made a brave effort to go back and see to the rear gunner but was beaten back by the heat. Later, with Major Braithwaite, who was uninjured, Downton was captured. Once the story of his actions was known he was recommended for the CGM.

For over two years Downton remained a prisoner of war, at first in Italy and then in Germany. His release finally came in May 1945 and he returned to his native Australia, arriving home that August. He was released from the RAAF in October.

*

Harold Vertican was seventeen years old when he joined the RAF. He served an apprenticeship as a metal rigger for three years and then went through a year's conversion course as an engine fitter.

In 1938 he was selected for a pilot's course which he successfully completed which brought him promotion to sergeant and a posting to Iraq where he was engaged on ferrying duties between Iraq,

Persia, Egypt and the Sudan, with 70 Squadron.

He returned to England in 1941 to become an instructor with the rank of flight sergeant but in 1942 returned to the Middle East and completed thirty ops on Wellingtons with 148 Squadron based at Gardabia Main to begin a second tour, this time flying four-engined Halifaxes.

His main operations were flown over North Africa, Sicily, Crete and Greece. On 2nd May 1943, he was recommended for the DFC having completed forty missions. On one occasion he flew over 700 miles back to base on one engine. After a six-hour struggle he found his base covered in fog so had to fly to another base where he landed, having been flying on one engine for $9\frac{1}{2}$ hours (this was on his Wellington tour).

On the night of 6/7th May, he was detailed to attack enemy troop and motor transport concentrations on roads leading to Tunis; his Halifax was DT501. They were airborne at 10.56 pm and after flying for $2\frac{1}{2}$ hours and while over the Sousse region, the starboard engine failed. Vertican decided the mission should be aborted. He flew the bomber out to sea to jettison the bomb load, then set course for an emergency landing ground. At 2.20, while at 8,000 feet, the port inner engine also failed. Warrant Officer Vertican continued to fly on two engines for a further $2\frac{1}{2}$ hours with only his PA compass, altimeter and air speed indicator working. He ordered the crew to jettison everything that was moveable and anything that could be hacked away from its fittings with an axe.

Despite these efforts, they still lost height and unknown to the navigator, Flying Officer Tempest, the wind had altered in strength and changed direction. They were thus being blown out to sea and away from their intended course.

At 4.45 am, the starboard outer engine failed and Vertican was forced to turn into the wind and try for a sea landing. He ordered the crew into ditching positions and ignoring the rough running sea, made a very good ditching some seventy-odd miles north-west of Tripoli. The Halifax stayed afloat and the dinghy was released and inflated. All clambered in but only two bottles of water were saved as the others, carried by hand, were dropped during the ditching and washed to the rear of the aircraft by the inrush of water. The Very pistol was also lost. The aeroplane remained afloat for about seventy-five minutes. Everyone was sea sick in the rough sea and it was not until the 8th that anyone wanted to eat or drink. A rationing system was instigated by Vertican which consisted of chocolate, Horlicks tablets, barley sugar and a little water.

When they became overdue, a number of pilots were sent out to search for them, flying Wellingtons, as Halifax aircraft could not be spared. Nothing was seen and so the CO, Wing Commander Warner, and Flight Lieutenant Craig, visited Castel Benito and Tripoli to investigate reports made about the operations of the 6/7th.

Meanwhile, in the dinghy, Vertican and his crew were soaked when it rained heavily on the third day, and it became very cold that night, but at least they had a good drink of water. During the daylight hours it became so hot that the men took it in turns to take a swim to cool off.

On the eleventh day they saw the coastline and soon the dinghy got into the shore breakers. Sergeants Curnow and Gordon, bravely jumped into the sea and helped guide the dinghy through the surf. Everybody then jumped ashore at 2.50 pm. They discovered they were about twenty-five miles west of Homs in Tripolitania. This information came from an Arab who appeared and offered help. They asked for water which was brought from a nearby village in the sand dunes.

Flying Officer Tempest gave the Arab a message to take to the nearest military base, for medical aid. The village turned out to be Absolam Ben Sin and the Arabs there were of the Oudoul tribe. They were given food, fourteen raw eggs, two loaves of bread, some goat's milk, dates, tea, coffee and cigarettes, plus welcome shelter. About 9 pm a Soudanese patrol arrived but none of them could speak English. They only had camels and donkeys, so in sign language they asked for motor transport. That night, the men were forced out of the shelter by fleas, so spent the night in the dinghy which had been brought ashore.

On the twelfth day another Soudanese patrol arrived, this time with two lorries and so they began their journey through the dunes, often having to stop in order to dig themselves out of the soft sand. They arrived at Homs and safety, on 18th May. One month later, Vertican was recommended for the CGM. They had been lucky.

Without the help of the Arabs they would not have made it. Harold Vertican, known as 'Mickey' to his friends, went on to become a test pilot in 1943. The next year he was commissioned and returned to England on radar experimental flying duties. After the war he stayed in the RAF, retiring in 1964 as a flight lieutenant.

Fighter-Bomber Pilot

These were the only CGMs given for specific actions in the Middle East. It was to be another year before anyone of the Desert Air Force received another by which time it was operating in Italy.

On 27th May 1944, Flight Sergeant John Casson, from Salisbury, Rhodesia, was with 250 (Sudan) Squadron and based at San Angelo, Italy, flying P40 fighter bombers. On this day he was detailed for an armed reconnaissance mission along the Caprino-Arche-Frosinone road. Near Frosinone, twenty-plus enemy motor transport vehicles were spotted and Red Section, led by Casson, in Kittyhawk FX761, went into the attack. He scored four hits on the road and claimed one motor vehicle destroyed when it blew up with flames shooting up to 300 feet. The section then attacked in pairs and obtained further hits. Blue Section also attacked and scored further hits; it was believed the transports were carrying vital petrol.

The flak was intense from several ground positions. Casson had made his two runs in the face of severe opposition. In his second attack his Kittyhawk was hit by a shell and he was seriously wounded in the thigh and his aircraft holed and a tyre punctured. Although bleeding profusely and almost passing out through pain and loss of blood, he flew his aircraft back to base, a distance of fifty miles. Unable to operate one rudder bar because of his wound and with a shredded tyre, he nevertheless made a good landing. Only when being lifted out of his cockpit did he finally pass out. He was recommended for the CGM the same day by his CO, Major J.R.R. Wells DFC, but sadly he succumbed to his wounds.

John Casson was buried on 28th May in British Cemetery No 12. His action came at the height of the Battle for Rome and apart from Casson, the squadron got quite a mauling in this operation. Wingman to Casson, Sergeant Barrow, did not get home. Flight Lieutenant McBryde, an Australian, had to force land but luckily, inside Allied lines. Four other aircraft came back with holes shot

(*Right*) Flight Sergeant Vertican CGM, DFC (now commissioned)

(*Below*) Kittyhawk in Italy

through them. At the time of his death, Casson had flown ninety-eight sorties and had 496 flying hours in his log book.

Number 39 Beaufighter Squadron had been operating in the Middle East since 1940 with a variety of aircraft – Blenheims, Marylands, torpedo-carrying Beauforts, before converting to Beaufighter Xs in June 1943. It had operated on all types of sorties against Axis shipping over the Mediterranean from both North Africa and Malta. By the summer of 1944, 39 Squadron was flying as part of 328 Wing, flying from Italy out over the Adriatic and across to Yugoslavia.

Flight Sergeant Stuart Somerville Campbell was with 39 Squadron. It was based at Biferno, Italy, and when he was recommended for the CGM on 4th August, he had flown on twenty-three operations as a navigator. Nine of these were long range fighter patrols, ten anti-shipping strikes and four daylight intruder patrols.

On 9th June, flying with his regular pilot, Flight Sergeant Kyrke-Smith, on a night operation to the Spezia-Leghorn area, their aircraft was hit by flak. This started a fire in his compartment in the aircraft, wounding him in the leg. Yet he fought the fire while his pilot continued to attack a 500-ton motor vessel. The fire destroyed his log, radio and VHF equipment. Despite his wound and some slight burns, plus the loss of his equipment, Campbell navigated them back to base at Alghero where a crash landing was made in the early hours of the morning.

On 23rd July, their aircraft was hit by flak when flying over Greece and a similar fire started as in the June sortie. He put the fire out, and again, without radio, they got back to make a crash landing in a field near base.

After knowing of his CGM award, he continued on ops and on 7th September, in Beaufighter LX877 'K', they set out to attack landing barges. They took off at 6.30 pm and arrived over the target area on time, only to be met by a terrific concentration of light gunfire. Their aircraft was hit in both engines and they were last heard over the radio trying to make the island of Vis on the Yugoslav coast. By the 8th there was still no news of them but the next day a message came from Vis to say that Kyrke-Smith had been picked up by a partisan ship and landed on the island, though there was no news of Campbell. On the 10th, Flight Lieutenant Payne flew across to pick up the downed pilot.

Apparently Smith had made a ditching south of Brac Island. Campbell was injured and despite several attempts to release him from the cockpit, Smith failed to get him out before being dragged down by the sinking aircraft.

Campbell was described in the squadron records as being one of the oldest navigators on the squadron, but his exact age is not recorded. He has no known grave but is remembered on the Malta Memorial, Panel 13, column 2.

*

The final CGM in this theatre of operations went again to a 250 Squadron pilot on fighter-bomber missions. It came in fact on one of the last missions undertaken by the squadron in Italy; by that date, 30th April 1945, the war was almost over.

The target was as ever, motor transport, this time in the Udine area, in northern Italy. A convoy of retreating enemy vehicles was making a dash for the Alpine passes under cover of bad weather. Once the six Kittyhawks were spotted they came under intense flak and small arms fire from below.

Flight Sergeant Dennis Evans was flying Kittyhawk FX568, and early in the attack he was hit by this ground fire but remained in his dive and pressed home his low level run, destroying two vehicles. With his aircraft badly damaged he headed towards the Allied lines but with smoke and fumes choking him, he was forced to bale out. He came down in enemy territory, evaded capture and returned to his squadron three days later, with the help of local partisans.

On 22nd May he was recommended for the CGM. In this was mentioned other operations in which he had participated between 9th and 30th April. In this period he had flown twenty-six close support ops for the Eighth Army's offensive. On the 11th he took part in an attack on an enemy intelligence HQ in the village of Fantazza. During this attack his aircraft was hit in the fuselage fuel tank. Undeterred, Evans made the attack and completely demolished two buildings with his bombs.

CHAPTER TEN

Build Up To Invasion

No sooner had the Battle of Berlin ended with the advent of spring in 1944, than Bomber Command continued its attacks on other parts of Germany. Soon, however, Sir Arthur Harris had to begin diverting much of his night effort in support of the coming invasion of Europe.

However, at the end of March came a raid that went down in history not because of the target, nor because of the success – but because of the losses.

Nuremberg

A force of 795 RAF heavy bombers headed out to distant Nuremberg. Even though it was a moon period the weather forecast was for cloud over the route and clear over the target. In the event the route was clear and the target cloud covered. German night fighters had a field day. Ninety-five bombers failed to return. One pilot, Pilot Officer Cyril Barton of 578 Squadron, was awarded the Victoria Cross – posthumously – on this night, and another airman won the CGM.

Sergeant Leslie Chapman was flying as a wireless operator in Lancaster R5856 of 61 Squadron on his third operation since joining the squadron. His pilot, Pilot Officer D.C. Freeman, lifted their Lancaster off the ground at 9.58 pm. Just north of Frankfurt they were attacked by three fighters (two Ju88s and one Me110) and suffered damage to the fuselage. The main hatch was blown off, the upper turret holed, flaps riddled, starboard wing holed and the front windscreen and astro-dome smashed. The crew had heard the gunners blazing away and also a warning yell, then a terrific explosion.

Sergeant Chapman was hit in the back, neck and head while three others of the crew, the navigator, flight engineer and upper gunner, were also wounded. Freeman made the decision that they could not go on to the target and after jettisoning his bombs, set a course for base. He was sitting in a gale of cold air that blew through the smashed windscreen. With the navigator, Sergeant Thomas, badly injured, it was left to the bomb aimer, Sergeant D.G. Patfield to navigate. Only moments before Patfield had been surrounded by small flames and a sticky, wet face, which he thought was blood, but turned out to be hydraulic fluid from the front gun turret. Most of the turret had simply disappeared and the severed pipes were spewing out the oil. He could see too that one engine had been hit.

With only a torn map and a pencil left of the navigational aids, Patfield was going to have a difficult task. He worked out a course but then passed out. His oxygen mask had been damaged in the attack and he spent the rest of the trip unconscious under the navigator's table.

Sergeant Chapman, disregarding his wounds, tried to obtain a fix on his wireless, but owing to the distance from base, it was over half an hour before he managed to do so – they were still a hundred miles inside enemy territory. Chapman obtained another fix, then went forward and plotted it on the map. Returning to his set, he obtained a further five fixes.

With these, Freeman was able to reach England and make a safe landing, and it was only then that Chapman revealed that he too had been wounded. He was awarded the CGM and Freeman the DFC. Both were Lincolnshire men, Chapman coming from a village near Spalding, Freeman from Gainsborough.

The crew never flew together again and, unfortunately, Freeman and the two gunners were killed in later raids. Leslie Chapman, too, returned to operations, but on 1st February 1945 was killed in a crash. He and his new crew had been detailed for an operation to Siegen, his pilot being the 'A' Flight commander, Squadron Leader Horsley. As they took off at 3.42 pm, the port outer engine was seen to fail. The aircraft made a tight circuit and belly landed on the base at Skellingthorpe where it exploded and caught fire. It was later revealed that the bomber was overloaded by some 166 lbs. All the crew, including Flight Sergeant L. Chapman CGM, were killed, with the exception of the rear gunner, Sergeant Hoskinson. Chapman is buried in Whaplode Cemetery, Lincolnshire.

*

As the build-up for the invasion began, Bomber Command began to divert some of its mighty effort to a variety of assigned tasks. On the night of 10/11th April, 90 Squadron flew a special operation to drop supplies to the French Resistance fighters.

Sergeant Edward Dyson Durrans was born in 1921 at Batley, Yorkshire, and enlisted in the RAF in 1941. After a wireless operator's course he was posted to the army who at this time were short of trained wireless men and he was attached to the 55th Signals Division, at Melbourne, Yorkshire. He retained his RAF uniform, of course, and wore in his 'fore and aft' cap, the white flash of trainee aircrew. Because of this he had to endure a few choice comments from the 'brown jobs'.

Eventually, after travelling the length and breadth of England with the army, he returned to his course and then onto OTU and conversion unit before a posting to 90 Squadron in December 1943.

On 10th April he and his crew took off at 10.59 pm in Stirling EF182. The supply drop was successful but on the return trip they were hit by anti-aircraft fire. Sergeant Durrans was wounded in the back and thigh, and had one leg broken by shrapnel. Despite the loss of blood he remained at his set and continued to send out and receive messages until they reached the safety of the English coast.

Arriving back over England the pilot, Warrant Officer Field, made a crash landing at Friston near Beachy Head, and Durrans was taken to Eastbourne hospital. In the next bed was Sergeant Waller, the flight engineer, who had also been wounded, but only a few hours later he died.

Two days after his arrival, Sergeant Durrans was taken to the RAF hospital at Halton. Here he was told, on the 25th, that he had been awarded the CGM and on the 26th he received a telegram from Sir Arthur Harris, congratulating him on his decoration. In all he was to spend three years in hospitals, undergoing fourteen operations in which he had skin and bone grafts on his leg. It was here that he met his future wife, who was a WAAF at RAF Halton.

On leaving the RAF in 1947, Ed Durrans went back to his old job

(*Right*) Flight Sergeant L. Chapman CGM

(*Below*) Buckingham Palace – Flight Sergeant Durrans

as a draughtsman but in 1949 he once again broke his leg and as a result he now has to wear a full length caliper of steel on his leg.

Although awarded his CGM in April 1944, he had to wait till 5th March 1945 before he was well enough to attend an investiture at Buckingham Palace, and then he was on crutches. During the ceremony he was photographed while receiving his medal by the King, and this appeared in *The Times* the next day, and *Flight* magazine on the 15th. This had been the first occasion members of the press or cameramen had been allowed inside the Palace during one of these occasions. In the same investiture, Major Tasker Watkins, now an eminent judge, was presented with the Victoria Cross.

*

The story behind the CGM awarded to Sergeant Peter Engbrecht, a Canadian from Whitewater, Manitoba, will show that an air gunner's life was anything but easy.

In 1944 he was serving with 424 (Tiger) Squadron RCAF, at Skipton on Swale, Yorkshire. Most of No 6 Canadian Group squadrons were based in Yorkshire and 424 were part of that Group. With D-Day looming, attacks on German troops and supply lines were vital. One such target, on 27/28th May 1944, was Bourg-Léopold, where there was a build-up of German troops. His pilot was an American, Flight Officer J.C. Keys USAAF.

They successfully bombed the target but they were attacked, not once but fourteen times by fighters. During the attacks, the rear guns became u/s, so Sergeant Engbrecht in the upper turret became the Halifax's only defence. In this role he more than held his own. He spotted one single-seater fighter, aimed his guns, fired, and saw it go down in flames. This was confirmed by the pilot, flight engineer and bomb aimer.

They were then attacked by a FW190 but by now he had only one gun working, yet by firing with extreme accuracy he saw the Focke Wulf break away and go down in flames. When his second gun failed, he continued to give Keys correct evasive action to avoid further fighters.

On another operation on 10th June, four days after D-Day, they were returning from an operation to Versailles when their aircraft was hit in an attack by a Messerschmitt 109, but he was more than equal to it and shot it down, followed by the destruction of a Me110.

He was recommended for the CGM on 27th June, and his medal was presented to him at Skipton on 11th August by the King. He celebrated the next night during a raid on Brunswick when he and the rear gunner, Sergeant C. Gillanders, shot down two more night fighters, a Me110 and a Me109, for which Gillanders received the DFM. This team repeated this action over the Falaise on 14th September with another Me109 and a Me410, and Gillanders claimed a probable victory four nights later.

After only three operations, Sergeant Fielder Bennett Dew, of 78 Squadron was also over Bourg-Léopold on 27/28th May. His aircraft, Halifax LW519, took off at 12.16 am, the skipper being Flight Sergeant Long, a Canadian.

They bombed the target at 2.15 am from 11,000 feet, and the attack seemed quite concentrated but when back over the French coast they were attacked by a fighter on three occasions. This started extensive fires in the bomb bay and fuselage. The fuel supply to both starboard engines was cut and Sergeant Dew was wounded in the foot, thigh and arms. Two other members of the crew were also wounded and the bomb aimer, Sergeant Bell, baled out.

Sergeant Dew, ignoring his wounds and blood loss, managed to extinguish the fires even in his weakened state, then got one of the engines going and supplied with petrol. He then collapsed and for the rest of the trip home was unable to move, but continued to direct other crew members in the flight engineer's duties. On landing at RAF Woodbridge, Dew was taken to hospital where his right foot could not be saved and it was amputated. He was recommended for the CGM, four days later.

The Battle for Normandy

The second night after D-Day, 7/8th June, Sergeant Gilbert Ebenezer James Steere was detailed as flight engineer with Squadron Leader W.B. 'Andy' Anderson's (Canadian) crew, to attack the marshalling yards at Achères, near Paris. In Halifax LW128 of 429 (Bison) Squadron, RCAF, they took off at 11.15 pm.

Near Dieppe on the outward journey they were hit by flak, much to Steere's surprise, as this was supposed to be an easy target. Being hit by flak was not a new experience for him and the crew. During eighteen operations, they had had three different aircraft owing to flak damage. The worst trip he remembers was a raid on Karlsruhe in April 1944. On that occasion a large piece of shrapnel came through the aircraft, only missing him by the fact that it lodged in the wood of the rest seat. Even so he was thrown across the aircraft and his oxygen mask ripped off. He fell unconscious and had to be revived by Warrant Officer Benning, the wireless operator. He will always recall the sharp explosion, ripping metal and the checks he had to carry out as a flight engineer, because of the damage, upon their return.

After his initial surprise over Dieppe, he heard a call from the pilot to jettison the bombs and then Anderson gave the bale-out alert. There was an unusual urgency in the skipper's voice. There must be a serious problem for Anderson had never given that order before when they had been hit. Unknown to most of the crew was that Anderson had been mortally wounded.

The drill for abandoning the aircraft, was for the navigator, wireless operator and bomb aimer to go out of the front hatch in the nose, the gunners out the back hatch and rear turret. The engineer would go past the pilot, down three steps and then also out of the front hatch. Putting on their parachutes, the first three men went out the front hatch. Gilbert Steere then remembers:

I then realised something which changed the whole course of events and to a large extent changed my life. I saw that Andy, the pilot, had passed out. He was slumped against the side of the cockpit and the aircraft was beginning to nose down. The rear gunner, Gordon Ritchie, says it was a dive but I still think it was a nosing down. I leaned over and took hold of the stick. The aircraft responded and the nose came up so here, with all the instruments intact, was an aircraft without a pilot. To add to this, there was now no navigator or wireless operator.

As the nose came up, both gunners, Johnny Mangione and Gordon Ritchie, both Canadians, met at the rear hatch. I plugged in my intercom which they had also done, and a short conference took place. I told them we had an unconscious pilot, a good aircraft and that I would like to try and fly it home.

Sergeant Mangione DFM, Sergeant Steere, Sergeant Ritchie

Sergeant G. Steere (now commissioned) being presented with the CGM by the King

At some point in the situation the pilot came round and asked what was happening. He was told and then he managed to select the distress frequency on the radio. This probably saved their lives when recrossing the English coast, as the Allies were nervous of any unidentified aircraft flying around close to their invasion forces and support craft.

> I badly needed to be in the pilot's seat, [continues Steere] so both gunners came forward and got Andy out, got him over the main spar and onto the rest position. Before doing so they gave him an injection of morphine. In his flying gear he weighed over 200 pounds and it was very difficult to keep the aircraft straight and level while they got him out. I then sat in the seat but, with my chest 'chute on, could hardly see out of the cockpit window. It was only after being debriefed that I found out that the seat I had been sitting in was full of blood, but I was far too busy to realise this at the time. My main problem was finding out where we were as we were flying above a thin layer of cloud of about 8/10ths. I had by now made up my mind that the two gunners would put Andy out on a dead man's line or static line and bale him out on this.
>
> I began calling on the R/T every few minutes hoping to contact ground control. I then saw the lights of a field and I received a reply to my calls. By now my flying was horrific. We were going up and down by some 500 feet a time. To add to this I was trying to make Control understand that I was flying the aircraft and why I could not possibly land it. I told them of my intention of getting out the pilot and then the remainder of us would bale out. The gunners moved Andy to the escape hatch and I then gave the order to bale him out.

Gilbert Steere found out some months later that ground control had turned out the local home guard but when they found Anderson he was dead. He had been shot through the lungs. In the air, the two gunners prepared themselves for the jump, and then Ritchie came on the intercom:

'OK, Gil, I'm going to jump now.'

Steere replied, 'OK, good luck.'

The other gunner took a little time to go himself, but in the end it was go or stay with the erratic flying of Gilbert Steere. Finally he walked down the three steps and straight out of the front hatchway all in one go!

I was now on my own. The fuel situation was on my mind. I could not see the dials from the pilot's seat and I had been flying for over four hours, so I felt the quicker I got out the better. After setting the aircraft to fly straight and level, I sat on the edge of the hatchway and allowed myself to be sucked out instead of going out head first. When I landed I tucked my legs up instead of using them as a spring to break my fall.

I watched the aircraft go down in a curve and crash, bursting into flames. As I came down I thought, 'It's my wife's birthday.'

I came down near some cottages and an old man put his head out of a window. He would not listen to what I had to say and threatened to shoot me as I must be a German if I had come out of a crashed aircraft. With my 'chute over my arm I walked to the airfield which turned out to be Benson, where I broke all the rules by walking across the runway and up to the control tower.

On 19th June, Gil Steere was recommended for the CGM and Sergeants Mangione and Ritchie the DFM. Of this Steere says:

I was pinned to the wall with everyone congratulating me on my award. All I had done was try to get home with a perfectly good aircraft when I knew the pilot, Andy, had no hope of baling out.

Steere did fly a couple of more operations and was then commissioned and became the squadron's engineering officer. It was twenty years before he met Gordon Ritchie again. They talked non-stop for forty-eight hours. Steere still thinks the Karlsruhe trip was worse than his CGM operation but of course the difference was that on the latter occasion the pilot was badly wounded, which left them without a pilot.

A bomb aimer faced with a similar situation was Warrant Officer Alexander William Hurse, an Australian from Victoria. He was born on 19th January 1920 and joined the RAAF in January 1942.

He joined 75 New Zealand Squadron in November 1943 and had completed twenty-nine ops when they took off to bomb railway targets at Nantes in France, on 10/11th June 1944, flying Lancaster ME751.

After bombing the target the Lancaster was severely hit by what was thought to be a light anti-aircraft shell, which exploded in the cockpit: The pilot, a New Zealander, Pilot Officer McCardle, was severely wounded and the flight engineer, Sergeant Benfold, received superficial injuries.

Warrant Officer Hurse took over the controls and with the assistance of the navigator, Pilot Officer A.H.R. Zillwood (later Flight Lieutenant DFC, he flew the aircraft back to England, despite McCardle urging him to order the crew to bale out. Knowing this would leave McCardle alone in the machine, he ignored the instruction. Not only did he then fly the bomber back but managed to land it safely. He was recommended for the CGM on 16th June.

Hurse went on to fly with 186 and 467 Squadrons, returning to Australia in 1945.

Another member of 75 New Zealand Squadron to be awarded the CGM was Flight Sergeant David John Moriarty, from Wanganui. He was born on 24th October 1918 and had been a civil servant before joining the RNZAF in September 1940 to become a pilot.

He had not flown many ops when he was detailed to fly on an attack on the village of Cagny, to the east of Caen, near the beachhead, in support of the British 2nd Army, prior to Operation Goodwood – the Allied breakout from the Caen bridgehead. The village held a large troop concentration as well as armour.

He took off in Lancaster HK568 at 4.30 am on 18th July, and one minute after bombing the target, and flying at 7,500 feet, his aircraft was hit by an anti-aircraft shell which exploded in the cockpit. He was injured by flying splinters of perspex which caused lacerations to his face and left side of his scalp, and damaged his left eye. The flight engineer, Sergeant Scott, gave him first aid but he insisted on staying at the controls and flying back to base. He did so for over an hour, in great pain and with no sight from his left eye, but he did so in 'masterly fashion' and got them down safely.

Moriarty was awarded the CGM, commissioned and ended the war as a flying officer.

Polish Gallantry

On the same day, 18th July, one squadron was assigned to attack German strongholds at Emiéville, a few miles from Caen, where the Germans were putting up a stubborn and fierce rearguard action. This was No 300 (Polish) Squadron.

Sergeant Jozef Pialucha, a flight engineer in the squadron, took off in Lancaster JA922 BH-J, at 4.16 am from RAF Faldingworth, Lincolnshire, piloted by Flight Sergeant Z. Stepian. Pialucha had escaped from Poland in 1939 and eventually reached England to join the Polish section of the RAF and later the Polish Independent Air Force.

Reaching the target, Stepian held the bomber steady during the bomb run, but they were hit by flak which caused the bombing installation to become u/s, and they were unable to release their bombs. The rear gunner, Flight Sergeant M. Zentar, was at that moment, revolving his turret, searching the daylight sky for enemy fighters, and the flak blast blew the turret beyond its usual limits, ripping open the steel doors behind him. He was immediately sucked back and outwards, the only thing saving him was his left foot which jammed into the doorway. There, many thousands of feet above the French countryside, he hung out, head downwards.

Sergeant Pialucha and the mid-upper, Flight Sergeant Derewienko, went to his aid but were unable to drag him back in. His foot then began to slip out of his trapped shoe as the lace began to snap, so Derewienko grabbed Zentar's trousers which then began to tear and rip with the strain.

In desperation, Pialucha climbed right through the opening between the turret and the fuselage, outside the aircraft, and hanging on with one arm, his feet on a tiny ledge, was able to loop a length of rope around the gunner's body to prevent his falling. By now they were back over the Channel.

The rope was then made fast to the turret seat and Pialucha re-entered the aircraft and returned to his duties as flight engineer, to help the pilot get their damaged Lancaster back to England and home. They made for RAF Tangmere on the south coast and landed at 7.11 am.

As the aircraft came into land, Sergeant Zentar was forced to swing from side to side to avoid hitting his head on the runway, but he made it. He was finally rescued as the aircraft came to a halt, his only injuries being a little blood coming from his ears and mouth. They had landed with a full bomb load too, which added to his and the crew's general anxiety. Little wonder that Zentar was suffering from shock. For his gallant act, Pialucha was awarded a well deserved CGM. When he edged outside, in the gap between the turret and the fuselage, there was no way he could have worn his parachute pack, so he had had no chance of survival if he had fallen or been sucked out. Being from a country occupied by the Germans, Pialucha's award could not be published in the *London Gazette*, but it was approved on 28th August 1944.

Sadly, Pialucha, the only Polish airman to be awarded the CGM, did not survive the war. He had been posted to 1586 (Polish) (Special Duties) Squadron flying Halifax and Liberator aircraft, engaged in dropping arms and supplies to partisan units in Italy, to the Polish Home Army and other resistance fighters in Europe.

He flew on many missions to Greece, Yugoslavia and Czechoslovakia as well as Poland. On the night of 1st September 1944, his Halifax, BB389, was hit by flak over Sombor, and crashed just south-east of Belgrade. Only the navigator managed to bale out, the rest were killed. Pialucha is now buried in the Belgrade Cemetery.

A third action on 18/19th July 1944, netted a third CGM for Bomber Command. This time the target was one of the many V1 flying bomb sites that the RAF and the USAAF had been attacking with some determination for some months.

Sergeant William James Bailey was born in Sunderland in 1923 and had earned his living in the carpentry and joinery trade before joining up in 1942. After his engineer's course and RAF training, he met up with his new crew at a Heavy Conversion Unit. His pilot was Flight Sergeant Buchanan, an Australian from Queensland. On 13th February 1944, they joined 78 Squadron, at RAF Breighton, Yorkshire. By July they were well on their way to completing their tour of ops, having just two more to complete thirty raids. Buchanan had been commissioned in the meantime.

Operation 29 for them was an attack on a V1 site at Acquet, in France. They took off at 10.05 pm in Halifax MZ788 and successfully bombed the target from 12,000 feet with sixteen 500 lb bombs. On

(*Right*) Sergeant W.J. Bailey CGM

(*Below*) Sergeant W. Bailey (3rd from left) and crew

the return journey, while crossing the French coast, they were intercepted by a German night fighter and hit by cannon fire which caused an explosion in the flight engineer's compartment and started a fire. As the shell exploded, Bailey was hit in both legs.

With the Halifax ablaze, Buchanan gave the order to bale out and the navigator, Flying Officer Rayment, the bomb aimer, Flight Sergeant McKenzie and rear gunner Sergeant Hamer, all baled out over the sea. However, the wounded Bailey was unable to do so, not only because of his leg wounds but also because he could not get to his parachute owing to the fire. He called Buchanan, telling his position and that he was going to try and tackle the flames. He then grabbed the extinguisher and did indeed put out the fire although suffering burns to his face and hands. He then discovered that the mid-upper, Sergeant McCannon, was trapped in his turret, so set to and got him out.

Buchanan was still making for England and on reaching the coast, the wireless operator, Sergeant Rice, called up the nearest base – West Malling, a fighter station in Kent – but they replied that the runway would be too short for a Halifax. Yet by now they were left with few options – they were committed. In they came at West Malling but quickly realised that the controller had been right. They were going to run out of runway. The end came and the bomber went through a field and over an embankment before stopping, but they were down and the Halifax did not catch fire.

Bailey was taken to the sick bay and from there to Preston Hall hospital near Maidstone. He underwent an operation to remove shrapnel from his legs and had his burns treated. The fire had also reached Buchanan, and the Australian had also to be treated for injuries to his face and hands. Two of those who had baled out, Rayment and McKenzie, were never seen again.

On 1st August, Bailey was recommended for the CGM by his CO.

CHAPTER ELEVEN

By Night and Day

By the end of July 1944, 'Bomber' Harris, while still supporting the Allied Invasion Battle Front, was beginning again to hammer German industrial towns. In August, his Command returned to the night skies over Germany in strength. He had acknowledged that his crews had needed to support the Allied cause in the Normandy and French battle zones, but believed that Germany's defeat lay in the destruction of its home-based war machine.

On the night of 25/26th August, 412 heavy bombers of 1, 3, 6 and 8 PFF Groups attacked the Opel motor factory at Russelsheim.

Flight Sergeant Robert Burton Maxwell was born on 20th January 1924, in Toronto, Canada. He joined the RCAF in May 1942, straight from school, and became a pilot. He joined 428 (Ghost) Squadron RCAF in the summer of 1944 and flew his first trip as a second pilot, gaining experience before taking his own raw crew into action. On 25th August, he was on the battle order – his first op as skipper.

They headed out to Germany. When still forty minutes from the target, his Lancaster was hit by flak, fatally wounding the flight engineer, Sergeant P. Recabarren. The starboard outer engine was rendered useless. The electrical system was knocked out, Sergeant Feasby the wireless operator, also wounded, and Maxwell himself was hit in the leg. Flames came from the smashed engine, but Maxwell feathered it and the fire went out. Undeterred, the twenty-year-old Canadian, in the full knowledge that he had every right to abort and return home, pressed on to the target and bombed it at 1.10 am from 12,000 feet. This had to be done manually as the bombing circuit had been knocked out when the electrics were hit. He then brought his damaged bomber home,

(*Left*) Flight Sergeant R.B. Maxwell CGM
(now commissioned)

(*Below* Bob Maxwell today

without wireless aid, to make an emergency landing at Thorney Island, near Portsmouth, where Sergeant Recabarren died shortly afterwards.

Maxwell was recommended for the CGM on the 28th, and was later commissioned. He went on to complete thirty-seven operations, having gone to 405 (Vancouver) Squadron RCAF in 8 PFF Group, and with the rank of flying officer, received the DFC. He left the RCAF in 1947 returning to his studies to obtain a degree in Applied Science in Chemical Engineering at Toronto University in 1951.

Four nights after Maxwell's brave effort, Stettin was the target. Well into his tour, Flight Sergeant Anthony Clifford Cole of 622 Squadron, was detailed to fly his twenty-seventh op on this raid (29/30th August), as wireless operator in Lancaster PD219 GI-L.

He and his crew took off at 9.10 pm. When over the Kattegat they were attacked and damaged by a Heinkel 219 night fighter. Sergeant Nicholls in the rear turret saw it coming in from 600 yards and yelled at the pilot to fly a corkscrew to starboard. As the Australian pilot, Sergeant Nielson, did so, the fighter began firing, Nicholls replying. However, only one of his guns fired owing to the turret becoming u/s a little earlier from an hydraulic oil leak. The fighter broke away and was not seen again. It had nevertheless scored hits on the Lancaster, which knocked out the intercom, smashed the perspex in the upper turret and shot numerous holes in the fuselage. The upper gunner was bruised and cut by flying perspex, the navigator, another Canadian, Warrant Officer Farquharson, was seriously wounded, and the engineer, Sergeant Thorman, also injured.

Anthony Cole was hit in the thigh by a piece of shrapnel which opened up his leg, and bared it to the bone. Ignoring his injury, he quickly reported to Neilson that there was a fire in the area of his wireless set but helped to extinguish it, even throwing bits of burning aircraft through the damaged astro-dome. He then helped to get the wounded navigator to the rest bed, still not mentioning his own extremely serious and painful wound. He then set about trying to get a fix in order to get home, even more important with the navigator out of action. When they reached eventually the English coast – only ten miles off track – the pilot decided to land as soon as possible, which he did with success.

By now Cole was on the verge of collapse and was immediately taken off to hospital. The recommendation for his CGM was forwarded on 3rd September by Wing Commander Swales, the squadron CO.

In contrast to Anthony Cole, Sergeant William Cridge was only on his fourth operation on the night of 23/24th September, when the target was Neuss. He was a navigator with 166 Squadron and in Lancaster PD242 they took off from their Lincolnshire base at 6.40 pm.

They reached Neuss around 9 pm and on the bombing run, often the most critical period of a bomb raid, they were attacked by a Ju88 night fighter. The rear gunner, Sergeant Hallett, and the fighter opened fire simultaneously. In the exchange Hallett was killed and Sergeant Cridge, and the wireless operator were both wounded. Cridge was hit in the face, the left arm and body by an exploding cannon shell.

The Lancaster suffered great damage to the hydraulics, wireless, navigational aids and most of the pilot's instruments were rendered u/s. Both gun turrets were wrecked but Hallett's fire did take effect too, and some of the crew saw the Ju88 blow up close to their bomber.

Continuing on the run-up, the bomb doors had also caught some damage, and this, with their other damage, decided them to seek a 'last resort' target, so they dropped their bombs at 9.22 from 16,000 feet. However, the open bomb doors then failed to close again.

Sergeant Cridge was in great pain and poor shape. He was losing blood but remained at his post giving his pilot all the assistance he could. He managed to get one of his navigational aids working again and thereafter passed corrections of course to the skipper to ensure a safe journey back to base. At times he nearly collapsed, and his charts were so covered in his own blood as to be useless to work on.

They finally managed a landing at RAF Manston, and Cridge was put forward for the CGM on the 29th.

Daylight Raid

Many of Bomber Command's flight engineers in Canadian or Australian squadrons were from the RAF, as was the case of Sergeant Ernest William Knight of 432 (Leaside) Squadron RCAF. He and his crew were detailed to attack Wanne-Eickle, an oil plant, on 12th October 1944 – in daylight. It was his twenty-sixth operation.

They took off at 7.27 am that morning in Halifax NP738 QO-J, flown by a Canadian, Pilot Officer Britton. They made their attack at 10.21 from 19,000 feet, bombing on a smoke concentration as ordered by the raid's Master Bomber. It was shortly after this that they were hit by flak which damaged the intercom, and one of the petrol tanks. The port engines were also hit and damaged. In the aircraft, the bomb aimer, Flying Officer Todd, flying his thirty-fourth operation, was killed, and Sergeant Knight hit in the hand which was partly severed. In spite of this serious injury, Knight continued to work on the damaged engines and kept them going.

They managed to struggle back to England, but then crashed into trees half a mile from Woodbridge airfield. Right up until the crash, Ernest Knight stayed at his post, tending the engines like a mother her babies. The crew survived the crash landing, and Knight was recommended for the CGM on the 15th by the CO, Wing Commander MacDonald.

*

Flight Sergeant Jackson Chartis Cooke was a Canadian who was posted to 103 Squadron on 9th September 1944. On 31st October, having completed just nine operations, he was detailed to fly a mission to Cologne, flying his Lancaster off from their Lincolnshire base at 5.49 pm (in LL964).

During the run up to the target, they came under heavy, intense and accurate anti-aircraft fire. At the very moment the 'Bombs Gone' call came, they were hit and badly damaged. Cooke found the fuel tanks on the port and starboard sides had been seriously holed and fuel could be streaming out into the darkness. He also discovered the starboard rudder controls no longer functioned, having been severed.

Cooke headed the Lancaster back towards Allied territory knowing it was odds on that they would not have enough fuel to get back across the North Sea to England. The engineer then informed him that it was a dead certainty – they were almost out of fuel already and were not going to get much further.

In the vicinity of Namur Cooke instructed the crew to bale out while he still had control. He then got out of his seat to bale out himself, thinking that everyone had gone. Suddenly he saw the mid-upper gunner still in the Lancaster, having accidentally pulled the parachute's rip cord inside the aircraft. Cooke climbed back into his seat and instructed the gunner to use the spare 'chute. However, by this time they were too low to attempt to parachute out so quickly made preparations for a crash landing. Cooke selected a field in which to put the Lancaster down with flaps and undercarriage up. During the approach both outer engines coughed to a halt, but he still achieved an excellent forced landing with minimum damage to the bomber. Fortunate too that he had made the Allied lines and safety. He was recommended for the CGM on 5th November.

Sadly he did not live to receive his medal. On the 29th November, flying Lancaster PB465 GV-F, he took off at 11.45 for the city of Dortmund, but failed to return. He was flying his thirteenth operation!

A 550 Squadron Lancaster landed at RAF Manston having reported colliding with an aircraft lettered 'F', which was seen to have its rear turret, starboard fine and rudder on fire. The bomber appeared to go down in a straight dive. Flight Sergeant Cooke is reported as being buried in the Reichwald War Cemetery in Germany.

With only eight operations in his flying log book, Flight Sergeant Stanley William Walters of 44 Squadron was faced with a problem on the night of 1st November. He was flying as bomb aimer on a raid to Homberg, in an attempt to hit the Meerbeck oil plant. 226 Lancasters with a small force of Mosquitos made the attack, but target marking was scattered and only 159 crews even attempted to bomb. One Lancaster was lost – from 44 Squadron!

When still twenty miles from the target, the bomber in which Walters was flying was hit by flak – LM650 KM-T. Its entire cockpit area was shattered and the pilot, Flying Officer Haworth, killed outright, and the flight engineer, Sergeant Seiler, severely wounded.

As the bomber nosed down, Flight Sergeant Walters scrambled back from his nose position to the smashed cockpit and helped the navigator to remove the body of their dead pilot from his seat. Then, regardless of the fact that he had never flown a heavy bomber before, Walters took over the controls and got it back to level flight.

By this time the engines were beginning to play up, and he quickly discovered that the hydraulics had been damaged. Having got the Lancaster under reasonable control, his first thought was to keep it steady so that the crew could bale out. However, once in control he decided to try and fly it back to England. Better to bale out over home ground than to go down to captivity in enemy territory.

Out over the North Sea and clear of known shipping lanes he jettisoned the bombs only to find that with damaged hydraulics, they could not close the bomb doors. Undaunted he kept going and set a course for Kent, at the same time keeping up spirits of the crew while issuing instructions for the treatment of the wounded engineer.

By the time he reached the English coast he knew the aircraft was too badly damaged for him to even consider trying for a landing, so he ordered the crew to bale out. He then headed the Lancaster towards the sea, and away from any built-up area, and only then did he bale out himself. Before leaving he ordered the crew to attach a static line to the body of their dead pilot and send it down in a parachute. Unfortunately the wounded Seiler failed to drop successfully and was killed. The aircraft crashed in the south of England at 5.45 pm. For his cool bravery and skill, Flight Sergeant Walters was recommended for the CGM on 8th November.

The night after Walters' courageous act, 2nd November, Bomber Command went to Düsseldorf, with a large force of 992 aircraft. As it turned out, it was the Command's last major attack on this city. Nineteen aircraft failed to return, one being DV396, a Lancaster of 467 Squadron RAAF.

Sergeant Derrick John Allen was an air gunner with this squadron, in Flying Officer Leslie Landridge's crew. Since joining the squadron in September, Derrick Allen had flown eight operations, the ninth beginning on take-off at 4.31 pm.

Shortly after leaving the target they were attacked by a night fighter. Allen spotted the German and opened fire and succeeded in damaging it. However, it continued to close in and shot away more than half the port fin and rudder, also damaging the fuselage and control trims. Because of the damage, Landridge had great difficulty in controlling the aircraft.

Again the fighter attacked out of the dark sky and once again

Sergeant Allen opened fire. In this attack the port outer engine of the Lancaster was hit and set on fire, while once more the fuselage was slashed with cannon and machine gun fire. An attempt was made to put out the fire but they failed. It would only take seconds for the wing to burn through.

The Lancaster lost height and went into an uncontrolled dive. Landridge had finally to give the order to abandon their craft. The rear gunner, Flight Sergeant Lewin – not a regular member of their crew – was unable to open the doors of his turret, so Allen climbed down from his upper guns and went to his aid. With an axe he chopped away the two doors, disregarding the fact that the aircraft was ablaze and losing height rapidly. Having freed Lewin, Allen went forward to bale out. Just as he did so, the aircraft broke its back and fell in two pieces.

Allen found himself in mid-air. Fortunately he had already clipped on his chest parachute pack, so he quickly pulled the rip cord. Unknown to him they had dived to within 2,000 feet of the ground so he was soon down to earth, landing in a tree and hanging by his harness, seven miles south-west of Spa, in Belgium. On climbing down he was met, not by a German soldier as he could have expected, but a herd of cattle! Allen recalls:

> It was about 2 pm and there was a bright moon. I thought it best to get away from the nearby crash site which was burning well. I headed for an old outbuilding which was at the bottom of a garden of a bungalow. I eventually decided that all the shouting I could hear in the distance was American, so I was pleased to give myself up. I was reunited with the rest of the crew later that night (morning) and it was then that I learned of the deaths of Les and Bill, whose bodies were only a few yards away from my tree.

The survivors were flown back to England, via Brussels a few days later. After a 21-day leave, Allen was passed fit for flying and became part of another crew to finish his tour. He was recommended for the CGM on 14th January 1945.

He married in 1946 and now has three children and five grandchildren. He served for nine years in the Royal Observer Corps and four years in the Air Training Corps. On his fiftieth

(*Right*) Derrick Allen CGM
(ATC Officer)

(*Below*) Derrick Allen
(Standard Bearer)

birthday, his daughter Judith wrote a poem covering the events of his fateful night in November 1944. It was entitled 'A Tree Somewhere in Belgium'. The last line reads: 'Oh where would I be, were it not for that tree?'

2nd TAF, Mediums

Not all the bomber CGMs went to crew of heavy bombers. Second Tactical Air Force operated medium bombers, mainly on daylight raids. Once the Allied armies had advanced onto the continent, some of these medium squadrons operated from airfields in France, Belgium and Holland, in order to be nearer their targets. These targets were usually troop or transport concentrations, or attacking enemy lines of supply.

Number 180 Squadron, which flew twin-engined B25 Mitchells, were operating from Melsbroek in Belgium by the autumn of 1944. On 3rd December, the squadron were assigned to attack a cross-roads at Straelen and Kaldenkirchen.

Mitchells taking off in Belgium

In Mitchell FW209, was Flight Sergeant Frederick Tomkins, wireless operator to the crew of Flying Officer Iddon. They took off at 8.40 that morning and flew successfully to the target area. However, they were hit by flak, causing damage to the port engine, cabin and fuselage. Both the pilot and navigator, Warrant Officer Allan, were wounded and an electrical fire started amidships. Very cartridges caught fire in the cockpit filling it with smoke. The immense glare was so great that the pilot was unable to control the aircraft either visually or on instruments.

The order was given to bale out, and Flight Sergeant Wolff, the gunner, went out. Flight Sergeant Tomkins, seeing the navigator fighting the fire, and seeing the smoke starting to clear, went forward to try and help the pilot. He crawled over the bomb bay and into the second pilot's seat where he regained control. The pilot, suffering from shock and loss of blood, continued to give him instructions. With the fire out, they plotted a course to the nearest airfield in Allied territory. On reaching the airfield, Tomkins, despite never having flown an aircraft before, made a safe landing guided by Iddon, at the Dutch base of Eindhoven. On the 5th, Tomkins was recommended for the CGM by the squadron CO, Wing Commander K.J. Powell.

CHAPTER TWELVE

Above and beyond …

In the last year of the war ten men received the CGM for continuous acts of courage in air operations above and beyond the call of duty, rather than for one specific act. The first of these was in February 1944.

Warrant Officer George Wilfred Brook was born in 1922 at Dalby, Yorkshire, joining the RAF in 1940. By 1943 he was a pilot in 550 Squadron, flying Avro Lancasters. On 19th February 1944 he was recommended for the CGM having flown twenty-nine operations, of which ten had been on Berlin during the prolonged battle during that winter.

In postwar life, Wilf Brook first became a pilot with BOAC and then went into farming in the south of England. Sadly I have to record that Wilf departed from this life in November 1984.

Warrant Officer Wilfred George Bickley was with 617 Dambuster Squadron. He had enlisted in March 1936 as a fitter apprentice. Later he became a part time gunner, who in those days wore the flying badge known as the 'Flying Bullet'. For this he received the princely sum of an extra 3 pence a day on top of his pay.

He joined 613 Squadron in March 1940, flying Lysanders with Army Co-Operation Command, with the rank of temporary corporal, but a month later he became a temporary sergeant, continuing with the squadron until December 1941. He then went to 44 Squadron at RAF Waddington, flying Hampden and later Lancaster bombers. Finally he joined 617 Squadron in November 1943. With the rank of warrant officer he remained with 617 until May 1944. When he was recommended for the CGM on 19th April, he had flown seventy-one operations. With 617 he was front gunner with the CO and force leader, Wing Commander Leonard Cheshire

Time of take-off

Warrant Officer Wilf Brook CGM (now commissioned) and Flight Sergeant Herbert Donaldson CGM (now commissioned)

VC DSO DFC. In July 1944 he was given a commission and left the RAF in 1946.

Flight Sergeant Arthur Harrington Jefferies was flying with 550 Squadron but was shot down on the famous but disastrous Nuremberg Raid in March 1944. Although the German night fighters were responsible for most of the losses that night, Jefferies and his crew were in fact brought down by AA fire over Liège. He went very early in the attack, being the fourth RAF aircraft to be shot down. He and three of his crew died, the other three becoming prisoners. (Lancaster LM425 BQ-M.)

He had flown in raids during the Battle of the Ruhr, the Battle of Hamburg, and the Battle of Berlin – in fact he had flown on eight of the sixteen attacks on that city. His CGM was recommended on 17th March, just two weeks before his last raid, and at the time he had flown twenty-five operations, so was nearing the end of his tour. He is buried in the Heverlee War Cemetery, Belgium.

Pathfinder Gunners

On the night of 8/9th May 1944, Sergeant Victor Arthur Roe was flying as rear gunner on an operation to Haine-St Pierre, France. As they were approaching the target – marshalling yards and loco sheds – two Junkers 88 fighters approached but were successfully evaded. Always the better ploy, as firing one's guns only drew the enemy pilot's attention to your position. They then encountered a Me110 which made four successive attacks with cannon and machine-guns. Sergeant Roe returned the fire on each occasion despite his turret having been hit, eventually becoming unserviceable while he was covered in hydraulic oil. When the fourth attack was delivered, the 110 was seen to be on fire in one engine and was claimed as probably destroyed. Roe was slightly wounded in the right arm by a shell splinter, but carried on clearing stoppages between each attack.

For this action he won the Distinguished Flying Medal, having at that stage of his tour, flown nineteen ops.

He went on to complete his tour and then joined 35 Squadron of the Pathfinder Force at RAF Gravely. When he was recommended for the CGM on 18th January 1945, he had flown on no less than sixty-six ops, and been promoted to warrant officer. This award was gazetted in April, but by that date, Roe was on the missing list. He

(*Left*) Captain Brook BOAC

(*Below*) Wilf Bickley CGM today

and his crew failed to return from a raid on Chemnitz on the night of 5/6th March, one of twenty-two lost from the raid. He hailed from Norwich and was just twenty-one years old. He has no known grave, but his name is engraved on the Runnymede Memorial, Panel 269.

*

One of the men who were awarded the CGM for an exceptional amount of operational flying, and who stands out with his record, is Warrant Officer Leonard Eric Gosling, another member of the famous 617 Squadron.

He was born on 8th August 1921 and enlisted in the RAF in June 1940. He began service life as a radar operator but remustered to become an observer in February 1942. Before joining 617, he flew with 619 Squadron as part of the CO's crew – Wing Commander J.R. Jeudwine OBE DSO DFC, later Group Captain.

Gosling's first operation was to Leipzig with Jeudwine on 4th December 1943, and during the Battle of Berlin he went to the Big City on seven occasions. In all he completed twenty-six ops with 619, then thirty more with 617, including the attack on the *Tirpitz*, via Russia, on 16th September 1944.

With 617 Squadron as with 619, he flew with the squadron commander, Wing Commander J.B. 'Willie' Tait DSO and three bars, DFC and bar. In June 1945 he transferred to Transport Command. He had amassed a total of 865 flying hours when his flying ended. His CGM was recommended on 30th April 1945. He also took part in the final operation against the *Tirpitz* on 12th November, when Willie Tait scored a direct hit on the mighty battleship.

From November 1946 until October 1948, Gosling was with Skyways Limited, as navigating officer and then the Lancaster Aircraft Corporation with which he completed 170 sorties during the Berlin Airlift. In all he flew over 2,500 hours. He joined British European Airways as operations officer in 1952 and later became operations officer at Rome Airport. He died at the relatively young age of fifty-one on 22nd April 1972.

In the Canford Heath area of Poole, his name lives on; as with other famous RAF and Fleet Air Arm flyers, a road has been named after him. It is called Gosling Close.

Coastal and Bomber Command

Warrant Officer Sidney Charles Hopkins completed some 209 operations as an air gunner when he was put forward for the CGM in May 1945.

His first two operational tours of duty were flown with Coastal Command, his third with 214 Squadron of Bomber Command. He flew in a variety of aircraft, such as Sunderlands, Catalinas, Liberators, Wellingtons and Flying Fortresses; the latter were with 214 Squadron as part of 100 (Bomber Support) Group flying radio counter measures.

He had completed 1,647 day flying hours and 590 night hours. In his log book there are many interesting entries noted, such as: '*Altmark*', 'search, found ship off south of Norway', 'Sank one sub', 'intense AA fire – Dunkirk beach', 'attacked by a Ju88,' 'Bombing raid on Stuttgart, Hamburg, etc'. By the end of the war he had amassed a wealth of experience.

It seems that many of the long-serving CGM winners were warrant officers. Another to receive this honour was Warrant Officer Squire Nuttall, another gunner with the Pathfinders. In September 1944 he had been awarded the DFM having flown on fifty-three operations. When, on 19th May the following year – just as the war ended – he was recommended for the CGM he had flown an incredible ninety-one ops as air gunner in Bomber Command, all with 35 Squadron, one of the first two Pathfinder Squadrons formed in 1942.

The third Pathfinder gunner, also recommended for the CGM on 19th May 1945, was Flight Sergeant Soloman Joseph Harold Andrew, also of 35 Squadron. He had completed fifty-four ops when recommended for the DFM in September 1944. This total had grown to eighty-nine when his name was put forward for the CGM. Many of his operations were with the Pathfinders, and as part of the crew of a Master Bomber.

*

Warrant Officer Sidney James Tregunno was on his second tour when he was recommended for the CGM on 29th April 1945. He had

(*Right*) Len Gosling, British
European Airways

(*Below*) Len Gosling's CGM
and other medals

completed forty ops at this time. He too was an air gunner, flying in Halifax bombers with 51 Squadron out of RAF Snaith, Yorkshire.

He had taken part in such operations as Berlin, Milan, Peenemünde, Le Creusot, and Munich, as well as all three major attacks on Hamburg. On his very first op, on his first tour, he and his crew had to ditch in the North Sea after being attacked by an enemy fighter, which he 'probably destroyed'. They had the distinction of being the first crew rescued by the airborne lifeboat method. In a later encounter with enemy fighters during a raid on Nuremberg, they were attacked by three for over twenty minutes but finally managed to shake them off.

CHAPTER THIRTEEN

The Final Assault

Sir Arthur Harris's bomber force continued to hit Germany till the end of the war. In the last months he sent his massive, highly trained, finely tuned air squadrons to smash all resistance from Hitler's Third Reich. These men, flying now many daylight raids, with the Luftwaffe being almost a spent force, continued too to show just as much courage during these final operations.

The first act of courage, which resulted in the award of the CGM in the year of 1945, went to Flight Sergeant Rupert Perry Longley. He had flown just five ops, when as wireless operator of Lancaster LM187 HA-H of 218 Squadron, captained by Flying Officer Banton, they took off for Neuss, a railway junction within the German railway network. They took off at 3.57 pm, on 6th January.

When approaching the target there was a violent shock and Banton, for a moment, lost control. The port outer engine began to run away and could not be brought under control; later it caught fire. It was stopped by cutting its fuel supply, not just to this engine but the port inner too. Banton had now to fly the aircraft back to base, an aircraft that was fast becoming uncontrollable and was steadily losing height. When they were down to 4,000 feet, Banton gave the order to abandon aircraft.

Flight Sergeant Longley made his way forward to the escape hatch. Here he found Sergeant Sims, the flight engineer, with a problem. His parachute had accidentally opened in the aircraft and he was reluctant to jump with it. He asked Longley if he could strap himself to his back and both of them descend on the same 'chute. In view of the situation, and with little time in which to discuss it further, Longley agreed. Strapped together, they made their way to the hatch. On the way, Longley's 'chute caught on something and nearly opened too, but he managed to grab it and keep it almost intact.

Together they jumped and their venture and Longley's courage and unselfish act deserved a better fate. However, the jerk impact of an opening parachute on even the dead weight of one human body is tremendous, let alone two. As the 'chute cracked open, Sims was torn away. Longley landed in a tree, fell twenty-five feet to the ground, seriously injuring his back in doing so.

On 2nd March, he was recommended for the CGM.

Mid-Air Collision

The target on 1st February was Ludwigshafen. Flight Sergeant William Eugene Crabe, a Canadian, was flying as mid-upper gunner in Lancaster NG202 TC-T of 170 Squadron from RAF Hemswell. It was his eleventh operation.

Shortly after dropping their bombs at 7.24 pm from 16,000 feet, their aircraft collided with another machine – a constant danger and fear – badly damaging the tail of the Lancaster, smashing the rear turret and killing its gunner.

Flight Sergeant Crabe immediately left his turret and went to see if he could help his comrade, assisted by the wireless operator. Crabe cut away the side of the smashed turret and tying a rope around himself, climbed into the wrecked gun position. By now he was completely exposed to the slipstream and in danger of falling out, and of course, he was not wearing his parachute because of the confined space. However, his efforts were in vain as the gunner was already beyond help.

On 6th February, his bravery was rewarded with the recommendation for the CGM. Initially, the squadron commander, Wing Commander Hachbrook, put Crabe up for the DFM. However, at a later stage this was upgraded to that of the CGM by a higher authority.

*

On the Pathfinder Squadrons there were usually two navigators, one operating the H2S radar set, while the other undertook the orthodox navigating duties. It was in the former capacity that Flight Sergeant Thomas William Dennis Kelly of 7 Squadron PFF, was

Flight Sergeant W. Crabe CGM

Flight Sergeant Kelly CGM,
DFC

serving over the winter of 1944-45.

On the night of 13/14th February 1945, Bomber Command's main effort was an attack against Dresden by 796 Lancasters and nine Mosquito aircraft, a raid that itself became famous through controversy. However, on the same night, 368 aircraft – mainly Halifax bombers, attempted to attack the Braunkohle-Benzin synthetic oil plant at Bohlen, near Leipzig. On this night Kelly was part of the crew in Lancaster PB667 MG-Q, which was to act as Deputy Master Bomber.

They took off at 6.40 pm and over the target at 9.57 pm, dropped their Green Target Indicators, and the main force bombers were instructed by the Master Bomber to bomb short of them. By 10.04 a large area was covered in Green TIs. During the approach, Kelly and his crew had been engaged by very accurate flak which knocked out one engine and caused other damage to the bomber. Kelly himself was wounded severely in one leg, but realising the part he had to play, refused all offers of assistance.

They received permission from the Master Bomber to return to base but on the return flight still remained at his radar set, supplying the other navigator and the pilot, New Zealander Flight Lieutenant Phillips, with information. This helped with an accurate course and the avoidance of heavy flak areas. Kelly all the while was in tremendous pain but it was only when they reached the English coast that he relaxed and slipped into unconsciousness.

It was his thirty-fourth mission, so he was eleven off his total of forty-five, which constituted a PFF tour, and meant he could then be taken off ops permanently if he so requested. In main force squadrons the usual tour at this stage of the war was thirty. Of Kelly's thirty-four, twenty-one were flown as marker aircraft.

His recommendation shows that on 26th February, he was put forward for the DFM by Wing Commander Crackness, CO of 7 Squadron, but when it reached the Station Commander of RAF Oakington, Group Captain Binghamhall, he decided that the CGM was more appropriate. On the 27th, Air Vice Marshal Don Bennett agreed.

Kelly had already been recommended for the DFM on 21st January 1945, an award that was gazetted on 17th April.

Flight Sergeant George Baillie Ferguson was born in Victoria, Australia, on 10th May 1925, and enlisted into the RAAF in July

1943, which indicates that he was little more than seventeen years old at the time. He was trained as an air gunner in both Australia and England, and was awarded his 'AG' brevet in December and promoted from leading aircraftman to sergeant.

He was posted to 466 Squadron RAAF on 18th September 1944 and was with this unit until just after the end of the war. On 23rd February 1945, he flew his sixteenth operation, detailed to fly as upper gunner in Halifax NP968 HD-B, flown by Flying Officer Gilbert RAAF. The target was Essen in the Ruhr so they were sure to receive a warm welcome!

When approaching the target the flak opened up on them, dead on time and with great venom. The Halifax received hits and Ferguson reported he had been wounded but that it was not severe and did not need attention. However, Gilbert wanted to make certain so he sent the engineer, Sergeant Howard, back to take a look. On checking, Howard could not see Ferguson's face because of the face mask. The gunner still confirmed that he was all right, and that in any case, if he was relieved another person would have to stand the extreme cold as his turret was full of flak holes, cold air blasting in. He, of course, was wearing, as all gunners did, a heated suit.

So he stayed at his post, giving information of flak, etc, and then they were on their bombing run. They dropped their load successfully, then Gilbert turned for home. During the run home, Gilbert checked with his wounded gunner but did not receive a reply. This time the wireless operator, Pilot Officer Deuchard, went back to check. The reason the pilot had not heard a reply was because Ferguson's oxygen mask was clogged with congealed blood, blocking the microphone unit. He was ordered out of his turret and found to be severely wounded. A sliver of steel $2\frac{1}{2}$ inches long and $\frac{3}{4}$ of an inch thick had embedded itself in his cheek bone, causing a compound fracture of the upper jaw, knocked out several teeth, with subsequent bruising closing his right eye. His face generally was very swollen. When his face mask was removed his jaw completely collapsed and it was necessary to use the mask again as a support to keep his jaw in place, and to prevent further blood loss. So seriously was he injured that Gilbert decided to make a landing on the continent rather than return to Driffield, Yorkshire. They landed at Nivelles, one of the Duke of Wellington's old stamping grounds some 130 years earlier.

George Ferguson's gallantry and courage was described as one of the most heroic acts performed by any member of aircrew in the history of 466 Squadron. By staying in his turret, he played a great part in the safety of the aircraft from flak and fighters, for as mid-upper gunner he had a good all-round vision and was the main eyes of the aircraft.

He did return to 466 Squadron but not until 14th May, by which time the war had been won. On 1st October he returned to Australia to be finally discharged on 7th December 1945.

Bombed in Mid-Air

On the night of 23/24th February, the target was Pforzheim, a vital rail junction mid-way between Karlsruhe and Stuttgart, as well as the main rail line for the Germans, leading to the American 7th Army battle front. With 342 aircraft having pounded Essen that day (see above), mostly Halifaxes, it was a Lancaster force (367) that flew to Pforzheim.

Flight Sergeant John Bettany was a wireless operator in 625 Squadron, which he had joined in October, having flown seven ops with 576 Squadron. He had joined the RAF in 1941 after his sister's house in Coventry had been hit in the German air raid the previous November. He decided it was far better to give than to receive.

His pilot on 625 Squadron was Flying Officer Paige, and on this February evening they took off in Lancaster PB815 CF-O. The Master Bomber on this raid was Captain Edwin Swales DFC SAAF of 582 Squadron. His Lancaster was to be hit and badly damaged in the raid, but he continued to instruct the main force bombers. On the return trip he had to order his crew to bale out, and he died in the subsequent crash. For his heroism he received the Victoria Cross, the only South African Air Force officer to win this supreme award in World War II. Of the raid, John Bettany recalls:

When we commenced our bombing run, I took up position in the astro dome, and when looking up, saw another Lancaster approximately thirty feet overhead – bomb doors open! We had just bombed, but before I could give any warning the Lancaster, still directly overhead, released the lot. I saw the 'cookie' and other bombs go past but we collected a fair proportion of its incendiaries which struck us and stayed burning on the wings and fuselage.

I managed to get rid of quite a number through holes in the fuselage, but the aircraft was now burning quite steadily – then the fuel tanks started to explode. 'Basher' Paige decided it was time to go. The flames had destroyed the intercom and the lines to the gunners' turrets were u/s, so I was asked to pass the word to the gunners.

Whilst crawling over the main spa (no mean feat in flying kit) my 'chute got snagged on some wreckage and opened and started to smoulder, so having decided that it wouldn't be of any great use, I dumped it and carried on. I saw the gunners out through the rear door and then made my way to the front of the aircraft. I was very fortunate in that a spare 'chute we'd been carrying was still okay and in its sling near the navigator's table. I then had to return to the rear of the aircraft as the cockpit was ablaze.

Bettany baled out and landed safely as did the remainder of the crew. The navigator, Warrant Officer Sullivan, suffered a sprained ankle but that was the total sum of injuries. Bettany soon made contact with an American advance patrol in France and returned to Paris where he met the rest of the crew apart from Sullivan who was in hospital. He left Paris on 24th March, nearly a month after taking off for his sixteenth op.

Bettany rejoined 625 Squadron in time to take part in Operation Manna, the food dropping operations to civilians in Holland, and also Operation Exodus, the repatriation of released Allied Prisoners of War from Europe. He was recommended for the CGM on 1st March, his citation mentioning that he had thrown out fifteen burning incendiary bombs with his bare hands in an effort to save the aircraft.

*

On 18th March, Warrant Officer Angus Robb RAFVR was recommended for the CGM. He had completed forty-nine ops with 405 RCAF Squadron of the Pathfinder Force. On 7/8th March, his aircraft was detailed to bomb Dessau in Eastern Germany, Robb taking position in his mid-upper turret.

Just after they dropped their bombs, the aircraft was subjected to three attacks by Ju88 night fighters. The first 88 was shot down by

(*Above*) 625 Squadron 1944:
Flight Sergeant J. Bettany
CGM ringed

(*Left*) Warrant Officer K.J.
Dennis CGM

the rear gunner, and the second by Warrant Officer Robb. During the third attack, and then by flak, the aircraft was damaged, and Robb went back to the rear turret when it was known that the rear gunner, Warrant Officer Hainsworth DFM, had been hit. Robb, assisted Pilot Officer Van Metre, the wireless operator, in freeing the wounded gunner, who was also trapped, with flames all round him. They extinguished the fire and got Hainsworth out, though Hainsworth later died of his injuries. Van Metre then returned to his radio which he continued to operate despite burns to his hands, and Robb went back to his upper guns, now the sole defence of the bomber.

Van Metre was awarded the DFC while Robb received the CGM. Eighteen RAF bombers failed to return, but five German fighters were claimed as shot down, two by Robb and Hainsworth.

There are several interesting entries in Robb's log book, such as those noted after a raid on Essen – 'turret doors flew off over the North Sea', 'lost an engine over Amsterdam', and 'crash landed at Wattisham'. Then – 'Hannover, our last trip and our shakiest do. Bomb aimer switched on nose lights over the target.' He also mentioned the raid of 7/8th March, adding: ' ... third Ju88 was shot down by a Mossie. Coned at Dutch coast for $\frac{1}{2}$ an hour. Petrol tank holed.'

Angus Robb's medals now reside in the Bomber Command Museum at Hendon, and when the reorganisation is complete for the fiftieth anniversary of the forming of Bomber Command in 1986, one can expect to see his medals alongside others, all proudly displayed.

*

One wireless operator from down under was awarded the CGM in 1945 – Warrant Officer Kevin John Dennis. He was born in South Australia, at a place called Ceduna, on 5th September 1924 and he enlisted in the RAAF in November 1942.

Most of his early WOP/AG training was completed in Australia and when he arrived in England in January 1944, he had his half-wing up and been promoted to sergeant. After further training he was posted to 462 Squadron RAAF, in February 1945.

On 13th March, and on only his second operation, his aircraft,

Halifax RG384 Z5-M, skippered by fellow Aussie Pilot Officer Paltridge, was detailed to carry out a Special Duty Flight to the Frankfurt area. They were part of 100 Group, charged with carrying out radio counter measures with 'ABC' (Airborne Cigar) and 'Carpet' wireless transmission jamming. This was in fact the first operation that 462 Squadron carried out, using ABC and Carpet. They were integrated into the main force bomber stream carrying special wireless operators who spoke German, to pick up fighter controller and fighter pilot transmissions, then jam them.

Shortly after leaving the target area, they were hit and badly damaged by flak and the flight engineer, Sergeant Welsh, was killed and the navigator, Flight Lieutenant Shanahan, wounded. Warrant Officer Dennis was hit in both legs and most of one foot was severed. The bleeding could not be stopped despite first aid, but despite this terrible injury, he refused to leave his set and continued to do his duty until they made a crash landing at the emergency landing field at Juvincourt.

During the flight back, he received all messages sent and obtained a weather report for the area of the diversionary airfield which they were making for. He was flown back to England the next day. Dennis finally returned to Australia in January 1946 and demobbed in June, having meanwhile received the CGM.

On 23rd March 1945, Sergeant Jeffery George Wheeler was detailed to fly an operation to Bremen, to bomb the railway bridges there; it was to be his thirty-third raid with 101 Squadron.

He had joined the RAF in 1942 with the hope of becoming a pilot. To this end he went to Rhodesia for pilot training, but in 1943 he remustered to become a flight engineer, returning to England as such in 1944. Completing his training he was posted to a Heavy Conversion Unit at Bottesford and there he crewed up, before going to 101 Squadron at Ludford Magna. He and his crew began their tour that November.

The raid to Bremen would be flown in daylight. His pilot was Flight Lieutenant Cooke. The target area, they found, was heavily defended and their Lancaster was struck on the starboard side by flak. Sergeant Wheeler was badly wounded in the thigh, but said nothing; he just carried on with his duties until they were away from the target. Only then did he attract the attention of the navigator,

Sergeant J.G. Wheeler CGM

who dressed his wound. Despite being told by the navigator and the pilot to lie down in the rest area, he refused and continued to carry on with his job.

Both starboard engines had been slightly damaged over the target but he looked after them so well that the pilot was able to have full use of all four engines throughout the remainder of the flight home.

After landing at base he was taken to the hospital at RAF Rauceby to spend the next four months recovering. On 2nd April he was recommended for the CGM. He finally returned to the squadron in

September, remaining with them until April 1946. He was demobbed from the service in December.

Forced down in Sweden

Sergeant George Wilfred Simpson was an RAF flight engineer, but served in 463 Australian Squadron. He was well on in his tour of ops when he and his crew were detailed for a raid on 25/26th April 1945. The target was an oil refinery at Tonsberg in Southern Norway. It was Simpson's twenty-second mission and the one he would always remember.

They took off in Lancaster RA542, at 8.14 pm, with Flying Officer A. Cox in the pilot seat. On the outward route, just after crossing the Norwegian coast, about 150 miles from Tonsberg, they were attacked by a Ju88 fighter from below. In the course of its attack, hits were scored by the German on the nose of the Lancaster, shattering the frame and perspex of the bomb aimer's compartment, and destroying all the bombing equipment as well as some of Cox's instruments. It also tore gaping holes in the perspex around the pilot's and the engineer's sections of the aircraft. The bomb aimer was wounded and Simpson received serious shrapnel injuries to his left hand and shoulder, but he kept working with the pilot to control the aircraft during the action with the fighter. The Lancaster's gunners finally succeeded in shooting down their antagonist.

Flying Officer John Wainwright, the navigator, went forward and gave first aid to the bomb aimer. Simpson, as well as losing a good deal of blood from his shoulder wound, also suffered severe frost bite to both hands owing to the cold air blasting through the shattered perspex, and in endeavouring to shield the pilot from the cold slipstream.

As the bomb aimer had been badly hurt, George Simpson then went forward to help him, even though he now had only one good hand. He nearly fell through a gaping hole in the fuselage floor but just manage to save himself in the darkness. He then reported to the pilot that the bomb sight had been smashed and shot away.

Sergeant Simpson, despite the biting wind, managed to work the jettison bar and succeeded in getting rid of thirteen of their sixteen bombs, which fell harmlessly into the sea. Still in the blast of freezing air, Cox was getting weaker and weaker by the moment and so Simpson stayed by his side, helping as best he could. It was then they decided to land in neutral Sweden rather than try to reach England, so Cox headed the crippled Lancaster in that direction.

Locating a Swedish airfield, Cox tried to get them down. It took three attempts to land, Simpson helping by manipulating the throttles by facing aft. He was unable to use his injured hand which was bleeding and swollen, and his whole body was frozen by the raging slipstream that lanced through numerous holes in the cockpit, perspex and smashed in front of the bomber.

Cox's hands too were frozen, and he was now holding the control column with his arms, hugging it to him, as he came into land. As the wheels finally touched the Lancaster bounced 100 feet into the air but Cox maintained control. When they finally came to a standstill, Cox collapsed over the column and Simpson had to switch off the engines and petrol supply.

On 11th June, Simpson was recommended for the CGM, while Cox and Wainwright were both awarded DSOs. It was Wainwright's prompt first aid and care, which, on the later testimony of a surgeon in Sweden, saved the bomb aimer's life. They had all acted magnificently under extreme hardship, open to the elements and in great personal danger. In this they lived up to 463 Squadron's motto: 'Press On Regardless.'

CHAPTER FOURTEEN

Far East, Final Europe and Vietnam

This final chapter of heroic acts in a book of heroic acts covers a number of flyers whose stories in some ways are outside the neat categories of earlier chapters.

There were, for instance, three men who received the CGM for operations in the Far Eastern Theatre of the war.

Flight Sergeant Clarence Reginald Green was born 23rd April 1918 in Tweedvale, South Australia. He enlisted into the RAAF in April 1941, just after his twenty-third birthday and qualified as a pilot in December. Completing his training, he was posted to No 100 Squadron RAAF in October 1942. At first he served with this unit in Australia, and then in New Guinea.

On 1st December 1942, while on a reconnaissance, his Beaufort aircraft was attacked by three Japanese fighters. With expert airmanship he was able to evade the attacks and get himself into a position from which to repel the fighters. In the subsequent air battle his gunners were able to damage one fighter and to claim a second as 'probably destroyed'.

Then later, on 8th March 1943, whilst flying escort duty in the Buna area off New Guinea, he saw a merchant ship blow up and disappear after being attacked by a formation of nine enemy bombers, escorted by fighters. These fighters then went down to strafe the survivors in the water. Although his Beaufort was out-gunned and out numbered, he took on the fighters with such daring and skill that they broke off the engagement and flew off. He continued to patrol above the survivors until they were picked up, and by doing so, saved many lives. His CGM was gazetted on 10th March 1944, almost a year following his acts of courage.

On 31st May, in bad weather, Green was the only pilot to locate

Beaufighter Mk II Far East

and bomb the Gasmata Island airfields and supply dumps, off New Guinea. The following day he was notified of his commission, and in December was promoted to flying officer.

He left 100 Squadron in July 1943 and returned to Australia as an instructor, a post he held till April 1944, when he was posted to 8 RAAF Squadron. While serving with this unit he flew sixty-three strikes over enemy territory. He was promoted to flight lieutenant in January 1945 and received a Mention in Despatches in May 1945.

He returned to Australia in 1946 and left the RAAF in July 1947.

*

The second CGM for a Far East recipient was an unusual award, as the man recommended was a parachute instructor in the RAF. Flight Sergeant Thomas Emanuel White was attached to 357 Squadron, based at Dum Dum, flying Special Duties ops, dropping men and supplies to insurgent groups behind the Japanese lines, with Hudsons, Liberators and Catalinas.

On 14/15th March 1944, a Hudson aircraft took off at 11.30 pm to fly such a mission. At 3 am a signal was received that the Hudson had crashed. Four of the crew had been killed, and two seriously injured. The assistance of a doctor was urgently required.

At Dum Dum, the medical officer, Flight Lieutenant George Desmond Graham MBE, who had just joined the squadron, immediately volunteered to parachute into the area where the injured men were. A drop was organised, piloted by Flight Lieutenant James Albert King DFC, having flown an operation to the area only the previous night. Flight Sergeant White also volunteered to accompany the doctor, and to assist in identifying and burying the dead airmen.

On the 17th, a message was received that Graham and White had landed safely, but that one of the injured men had died two days earlier. The drop had gone well, and the two men had been met on the ground by Kokang guerillas dressed in blue uniforms under the control of Colonel Yang Yan Sang. Mule transport had been provided and after an hour's journey to the crash site, he found the wreckage half a mile west of the Po Ko village.

In a mountain hut, directly opposite the crash, the sole survivor, Flying Officer Prosser, the aircraft's navigator, was found. He had been given first aid by Major Leitch and Lieutenant Parsons of the American forces in the area. Prosser was found to have a fractured skull, cuts on his face and a fractured right ankle, plus a fever due to infection that had set in. Treatment was given, and he was attended by Leitch, Parsons and Flight Sergeant White.

After three days he began to improve, the men having kept watch on Prosser in relays around the clock, but then he had a relapse. Captain Hockman, an American medical officer then arrived, having journeyed from T Etang by mule, which had taken him five days. He and Graham tended Prosser constantly and he began to improve again. Meanwhile, the dead crewmen were all buried and given full military honours.

Then on 24th March, a message was received that a force of 200 Japanese soldiers was making its way towards their position, being even then only four hours' march away. Prosser was better but not well enough to be moved, so they decided to await further confirmation of the enemy's movements. This came on the 30th – the force was just thirty miles away and numbered 400!

On 1st April, they departed into China, Prosser being carried on a litter by twelve coolies. On the 8th the coolies vanished in the night and some hours were spent trying to engage new ones. In the meantime, White was sent ahead to see if he could recruit others. He returned on the 9th and they set off again.

Two days later they climbed a pass of 7,000 feet, then on the 12th they got a message to 357 Squadron. The next day they were transported in a weapons' carrier to Yunshin staying there at the 22nd Field Hospital. They then contacted the American Air Transport Command who promised to fly the party to India. They took off but then the aeroplane's undercarriage failed to retract, so had to land immediately. Prosser collapsed again and was air sick. They returned to Kunming where he was refreshed. They finally flew out on the 17th, arriving at Dum Dum at 6.30 pm.

Flight Lieutenant Graham praised the work of Flight Sergeant White, for the way in which he looked after Prosser, his help with the coolies and general all round assistance. They had covered over 100 miles, travelling for five days, and in all they had been away for a total of thirty-three days. White was recommended for the CGM on 17th May, and Graham the DSO. Flight Lieutenant King, who had flown them in, received a bar to his DFC.

Shot Down behind Japanese Lines

The third award for the Far East was to Flight Sergeant Derrick Teify Jones. He had enlisted into the RAF in 1942 although he originally volunteered the year before. Most of his pilot training was carried out in Canada, and on completion he was posted to 62 Squadron in Burma. The squadron was one of the essential supply dropping units, flying the DC3 Dakota aircraft over the jungles of Burma. His first operation was flown in November 1944 – a low level supply drop into the Kobar Valley.

On 5th May 1945, the war with Germany was just about over in Europe, but the Japanese war had another three months to run. On this day, Jones was detailed for an operation in support of the invasion of Rangoon, and they were to fly there via Ngoyah Bay. They took off at 6.15 am in a C47 – 'S' for Sugar. Over the village of Myaungnya all the supply aircraft came under fire and all were hit to some degree. The plan had been to fly in at 500 feet and make the supply drop to troops in the grounds of Government House. Jones's aircraft was hit and a fire started in the starboard main petrol tank which quickly became uncontrollable. Within seconds they had lost over 100 gallons of petrol and Jones had turned back towards friendly territory.

Then the starboard engine itself caught fire while the main blaze had now spread to the wing root. They jettisoned their supply load, and prepared for a possible forced landing. He could not, however, get the aircraft over the last range of hills between them and base so had to land. With some skill he brought the burning machine down in a paddy field south of the village of Alakwa. As soon as they stopped, the port engine caught fire and in the rush to get clear nothing was saved except one map. Two of the cargo handlers even left their shirts behind!

As Japanese troops were known to be in the area they ran for cover into the nearby jungle, but were soon overtaken by some Keren natives who had seen them come down. They indicated which direction they should travel, and the natives followed, carefully covering up their trail. They also brought them food and later found them a good hiding place. They also warned them that the local

(*Left*) Flight Sergeant D. Jones
CGM (now commissioned)

(*Below*) C47 supply dropping
in Burma

Burmese were not over friendly and had been known to co-operate with the Japanese. Only recently they had handed over the crew of a downed Beaufighter, who were promptly beheaded.

By now some Japanese soldiers had been organised into a search party and were out looking for them, but the Kerens gave them a direction opposite to the way Jones and his crew had taken. This journey entailed swimming two rivers but they eventually reached the village of Nyongong where they spent the night.

The next morning they were taken back into the jungle to avoid any passing Japanese soldiers or patrols and later the Kerens brought a Burmese to them who was a British agent, previously dropped by parachute into the area. The man had a radio transmitter in a nearby village, so they travelled there overnight. On the 7th a message was sent to say the crew was safe and asking for instructions.

Two days later a reply was received that a Catalina flying boat would pick them up if a rendezvous could be arranged. They agreed, and the Burmese agent and the Kerens said they would get them to Dansen Bay for a pick up at 7 am on the 11th.

They set off on the 9th, walked five miles, and then rested till the dawn of the 10th, eating some food they'd been given, mainly rice. At the village of Oncahuwg a party was put on for their benefit, as they were the first Europeans that had been seen since the war began.

At 7 am on the 11th they were at the rendezvous but nothing happened until about 10 am when a Royal Navy patrol boat was sighted. As it hove to, it sent a boat inshore to pick them up. A major in the Rescue Service paid the villagers for their help, and then took off Jones and his crew. On 13th May, Jones was put forward for the CGM.

At this time he had flown over 130 sorties of which sixty-three were supply landings and sixty-seven parachute drops. In the recommendation, mention was made of the fact that on the 5th, he had received burns on the face while trying to extinguish the fire, having handed over the controls to his second pilot.

*

There was one unique award of the Conspicuous Gallantry Medal to an American airman of the US 9th Air Force. On 16th July 1944, Sergeant James J. Reardon of the 558th Bomb Squadron, of the 387th

Bombardment Group, was recommended for this British award. He hailed from Hartford, Connecticut, USA, and had already been awarded the Purple Heart and Good Conduct Medal.

Just over a month previously, on 12th June, he was flying as bombardier during a raid over the Normandy battle front. It was a low level sortie and they came under concentrated ground fire. Just as they commenced the bombing run, the Marauders came under intense fire, and a burst of an exploding anti-aircraft shell smashed in the front of the aircraft, wounding Reardon.

None of the crew was aware of this and despite severe loss of blood, he stayed at his post and dropped his bombs on a concentration of enemy troops. They were then hit again by flak, a direct hit in the left engine, which threw the B26 out of control.

Heading north, the pilot, Captain Thomas J. James, struggled with the controls and they were out over the Channel before he managed to get the machine to fly in something like a normal manner. It was only then that Sergeant Reardon revealed his condition to the co-pilot, who then pulled him out of the shattered nose section to give him first aid. On landing he was rushed to hospital where he eventually recovered from his ordeal.

His award was not gazetted, but following recommendations and approval of both British and American officials, it was submitted for approval to the King in January 1945, and awarded shortly afterwards. Sergeant Reardon's unique award was in keeping with the high standard set by brother members of the Royal Air Force and recognised as such.

*

A British member of a medium bomber unit received the CGM following an action which took place in March 1945. Flight Sergeant James Mansfield Hall came from Jamaica, where his father was the Assistant Director of Medical Services. He had completed seventy-two operations with 180 Squadron, as an air gunner in their Mitchell bombers, when he was detailed to fly a mission to bomb the marshalling yards at Bocholt, on 21st March.

They took off at 9.15 that morning, in Mitchell HD386, from their Belgium base at Melsbroek, so their target was only 120 miles distant to the north-east. The pilot was Pilot Officer Dick Perkins.

The Allied armies were massing for the Rhine crossing but were still relying on the men of the RAF and USAAF to pave the way and make sure the enemy's lines of supply were disrupted so as to enable the crossing to be made with the minimum number of casualties. Everyone fully expected the crossing of the Rhine would be defended with all the energy the Germans could muster.

On arriving at the target they found the flak defences quite severe and the raiding force had a tough time. One aircraft was seen to suffer a direct hit whilst actually dropping its bombs, and exploded, while another Mitchell was seen with its port engine on fire and a large hole torn in its fuselage. Yet another B25 was hit and damaged, its wireless operator having one leg almost shot away. The pilot had to make an emergency landing.

Meanwhile, Perkins was taking their Mitchell into the bomb run. On and on they flew, flak exploding about them and Jim Hall began to wonder if their bombs were ever going to drop. Only moments before he'd seen the exploding Mitchell cartwheel down, taking with it Warrant Officer Roy Clipsham, a second tour man, who on this mission was carrying an RAF cameraman, Flying Officer Smith, to cover the attack. He was destined not to get back with his film.

Then the gun turret was hit. Hall lost consciousness and when he came to a few seconds later, the Mitchell was in a dive. He was choking with smoke that swirled about him and quickly extricated himself from the turret. As he did so he heard someone calling the pilot. When he had joined the RAF he had wanted to be a pilot but failed during his dual instruction days in Canada, having on one occasion failed to see a red Very light from the control tower as he came in to land.

When the flak hit them, they had just released the bombs, and Perkins had felt a heavy blow, and his right leg was knocked off the rudder pedal, just as the Mitchell began to dive. He fought to regain level flight but the controls felt like lead. He tried with his feet to get more leverage but found his right leg useless. There was a jagged hole in the cockpit where the flak shell had burst. One shell fragment had smashed through his right thigh and also entered his left leg, coming out the other side.

The first man to reach him was Jim Hall. Jim looked down and saw the blood pumping out of Perkins' leg at an alarming rate. As the Mitchell had dual control Hall clambered into the right hand seat and grabbed the control column. It had been eighteen months since he had last flown, apart from a few minutes' dual that Perkins had given him. However, Hall found the control column was useless, and it just flapped about in his hands.

Their only chance was to get Perkins out of the pilot seat. Hall and the navigator, Pilot Officer Roberton, managed to do this, and Hall sat down and took over the controls. The other gunner, Flying Officer Butler, tended to their wounded pilot, applying a tourniquet to his leg and giving him a shot of morphine.

The port engine was now giving problems and the intercom was dead, so on the VHF set, Hall sent out a 'Mayday' call but then that radio too became u/s. At that moment Hall saw an airstrip ahead, but it looked terribly small and made out of steel planking (PSP). It was British but obviously just a fighter strip as he could then see Spitfires dispersed on the ground.

Perkins began to talk Hall down. Looking at the instrument panel, Hall found he had no airspeed indicator, no rev counter, and no boost gauges working, and the port engine was still giving concern. To add to his problems, the hydraulics were found to be damaged and the bomb doors were hanging down. His main concern was losing vital flying speed as he came in, and stalling. They fired off two red distress flares, and flak helmets were put on Hall and the wounded Perkins.

Hall eased the Mitchell down gently, coming over the runway at 50 feet – then they struck the ground. Metal screeched on metal, tearing off the bomb doors. They bounced to 60 feet or so; Perkins shouted to him to turn off all the switches. Hall did so, avoiding the possibility of fire. Then the Mitchell was on its belly, finally sliding to a halt. No one was hurt in the crash and all were safe. Perkins' injuries kept him in hospital for the next two years but he survived, although he was left with a limp.

On the 22nd, Hall was recommended for the CGM and Perkins the DSO.

Glider Pilot at the Rhine Crossing

An award was made to the Army in May 1945, which was, in fact, the only CGM ever given to an army man, then or since. It was awarded during the assault on Germany, or as it became known, the Rhine River Crossing, north of Wesel.

Acting Squadron Sergeant Major Lawrence William Turnbull was serving with No 1 Wing of the Glider Pilot Regiment, which was part of the 1st British Airborne, under the overall command of the 6th Airborne Division. On 24th March, he was the senior pilot of a glider carrying a heavy load of personnel and medical equipment. Nearing the landing zone and while flying at a height of 2,000 feet, a loose tow rope smashed across the cockpit, destroying most of the glider's controls and breaking the machine's fin and rudder. His glider went down in a vertical dive, but by superb work on the remaining column, Turnbull managed to bring the glider back to level flight.

All the time the glider was under heavy small arms fire, together with a substantial amount of light 20 mm AA fire. The glider was hit at least five times, killing the second pilot and wounding two of the soldiers in the back. One of the wing tips was shot away and the fuselage holed and badly damaged.

Sergeant Major Turnbull remained calm and collected, and by sheer concentration brought the severely damaged glider and its load down to the landing zone only 100 yards from his assigned position. As the glider ground to a halt, the area was being swept by heavy small arms fire and the evacuation of the wounded was almost impossible. Nevertheless, Turnbull organised a stretcher party and, with complete disregard for his own safety, successfully extricated the wounded from the glider. Throughout the whole operation and especially during the difficult landing zone conditions, Turnbull showed the highest degree of courage and was an inspiration to others.

He was recommended for the CGM by his CO, Lieutenant-Colonel Murray, which was endorsed by Major-General Gale, commanding 1 Brigade of the Airborne Forces, and counter-signed by no less a man than Field Marshal Bernard Montgomery, C-in-C of the 21st Army Group. It was received on 14th May 1945 and passed on the 18th.

*

In April 1945, Warrant Officer Alan Penrose of 157 Squadron, was

(*Left*) Flight Sergeant James Hall CGM

(*Above*) Staff Sergeant L.W. Turnbull CGM (centre) at Buckingham Palace

(*Bottom*) Staff Sergeant Turnbull's medals

recommended for the CGM. He already held the DFC and bar for actions during two tours of fighter operations.

His first tour began with 125 (Night Fighter) Squadron, which at the time was part of 10 Group of Fighter Command. He was a navigator/radar operator. When he joined 125, in February 1942, they were still operating Boulton Paul Defiants, so he flew as an air gunner. Later in the year the squadron was re-equipped with Beaufighters and he teamed up with John Owen Mathews who had joined 125 in June 1941.

On 21st October 1942, in a Beau, Mathews spotted and gave chase to a Ju88 which they attacked and damaged. The only reason the combat was cut short was that his machine guns jammed. On 16th February the following year, flying from Fairwood Common, Penrose and Mathews again found and attacked a Ju88, but this time they were able to put claim to a probable.

In November they joined 157 Squadron, operating with Mosquito night fighters. At the time of the Invasion in June 1944, they became part of 100 Group flying night intruder operations and bomber support missions. On the night of 15/16th June, while in the Creil area, they attacked two Ju88s, claiming one destroyed and one damaged. The first was picked up on Penrose's radar at 12,000 feet. On closing, a visual was picked up and it was seen to be an 88. A one-second burst saw the port engine explode and half its wing fall away. Locating the second 88, they identified this, too, and a short burst from 75 yards was given. Hits were seen on the wings but then contact was lost.

On 12/13th July, while in the Étampes area, a contact was made while at 15,000 feet. Within fifteen minutes, during which time Penrose guided his pilot, a visual contact was obtained and once again it turned out to be a Ju88. A one-second burst was fired and an explosion occurred on the fuselage and starboard engine. The aircraft caught fire immediately and dived to starboard burning fiercely. It was then seen to explode on hitting the ground in a position ten miles south of Étampes. Their squadron was then involved in fighting the V1 flying bomb menace. By August, he and Flight Lieutenant Mathews had destroyed no fewer than six flying bombs. One they attacked exploded so close to them that the blast set the Mosquito's rudder and rear fuselage on fire.

On 14th August, he and Mathews were recommended for the DFC.

During a patrol on the night of 7/8th October, while on a High Level Support operation, a radar contact was made. Mathews reported:

> Contact was obtained at six miles, range, ten degrees above, crossing from port to starboard, doing a slow right hand orbit. Range was closed climbing from 8,000 feet to 9,000 and as we got a visual, 1,000 yards away, 40 degrees above, the enemy aircraft straightened out on a course of about 240 degrees, doing 160 ASI. We recognised it as a Me110 with long range tanks and opened fire with a short burst from 100 yards dead behind. Strikes were seen and a small explosion occurred in the starboard engine. Another burst and the starboard engine caught fire. The colours of the day were fired off and it dived down burning, hitting the ground with a very large explosion lighting the countryside for miles.

On 11/12th November, they were once again on a High Level Support Patrol, this time southwest of the Ruhr. Until the end of the bomber attack on Dortmund, they flew in the Cologne/Bonn area, keeping clear of the bomber stream. Then they made a contact in the Bonn area. It proved to be another Ju88 and two bursts from 150 yards scored strikes on the port engine and it was claimed as damaged. Then on 4/5th December, during a patrol north of the Ruhr, over the Dortmund airfield, a Ju88 was attacked and destroyed.

Two nights later on a patrol east of Frankfurt, a Ju88 was seen and Mathews attacked from fifty yards. The starboard inner petrol tank and the fuselage both caught fire. The night fighter dived steeply downwards, burning fiercely and was seen to crash amidst some houses in a town thought to be Limberg.

On the eve of Christmas 1944, a patrol three miles south-west of Cologne, gave support to a bomber attack on the city and Hangolar. A Ju88 was chased and shot down with a 1½ second burst, the starboard engine bursting into flames. As it turned to port, three parachutes came out in quick succession and three bodies sailed to

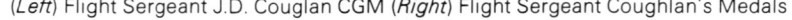

Mosquito XIX

(*Left*) Flight Sergeant J.D. Couglan CGM (*Right*) Flight Sergeant Coughlan's Medals

earth. The aircraft spun down in flames and crashed. Mathews and Penrose landed back at base at 9.43 pm with more than Christmas to celebrate.

A patrol to Karlsruhe on the 2/3rd January 1945 produced a Ju88 which they then lost, but then a second 88 was found ten miles north-east of Stuttgart. A two-second burst produced strikes and then a fire in the cockpit and central part of the fuselage. The 88 dived to port and exploded on the ground ten miles to the south-west of Crailsheim.

Penrose's second award came on 22nd January, when he was recommended for a bar to his DFC. All their combats were mentioned and his total ops had now reached 85. Mathews too received a bar to his DFC.

Their next encounter with the Luftwaffe was on 16/17th March. This time they were supporting a bomber attack on Würzburg and a contact was made in the bomber stream in the Stuttgart area. Then a Ju88 was identified and a burst of fire from 150 feet caused the port side of the fuselage to catch fire. It dived to port and exploded on the ground some twenty miles south of Würzburg.

Their final awards were put forward on 19th April, a DSO to Mathews and the CGM to Alan Penrose. By this time Penrose had flown ninety-seven sorties covering 350 hours. Fifty-seven of these operations had been in 100 Group on Support and Intruder missions. He had assisted in the destruction of some ten enemy aircraft and six flying bombs, plus the probable destruction of one aircraft and the damaging of three others.

Vietnam

There was a gap between the last award of the CGM of World War II and a post-war award. Corporal John Desmond Coughlan, from Sydney, Australia, was a member of the RAAF in the 1960s. He began a tour of operations in Vietnam with 9 (RAAF) Squadron in 1967, having been in the service for seven years.

Among his duties with 9 Squadron was that of a winch operator on helicopters. On 3rd October 1967, he was a crewman in a helicopter which responded to an emergency call from an American Army helicopter which had crashed in dense jungle deep in known enemy territory.

They flew to the scene and on arrival the helicopter was seen to be burning furiously. Corporal Coughlan volunteered to be winched down to search for and assist any survivors despite the fact that the crash had by this time undoubtedly attracted North Vietnamese and Vietcong forces. On reaching the ground he found a number of injured men. Aided by one of the less injured crewmen of the American helicopter crew, Coughlan located and prepared three of the most seriously hurt survivors for a winch lift. When this was completed he made another search of the area for other men. It was as well he did for he found more injured men from the 'chopper' and prepared them too for a winch lift by other helicopters which had been summoned and had now reached them.

During all this time, burning ammunition from the crashed machine was exploding and he was frequently forced to take cover as bullets struck trees about him. Nevertheless, all the injured were eventually got away and then he too was winched to safety.

On 13th January 1968, he was again winch operator when engaged in a night rescue of a crew and patients of a US medical evacuation helicopter which had crashed while evacuating wounded from an army company which was in action with enemy forces. As his pilot hovered at tree top level with its landing light on, their aircraft came under ground fire from the Vietnamese forces. They remained in this vulnerable and highly dangerous position for some time while the men on the ground were winched to safety. While this was going on, Coughlan had to lean well out of the helicopter as he issued instructions to his pilot.

His bravery and disregard for his own safety were recognised by the award of the CGM, which was later presented to him by the Governor-General of Australia, Sir Paul Hasluck, at an investiture in Varralumia, Canberra, in 1969, having been in the meantime, promoted to sergeant, and then to flight sergeant. He is still serving, with the rank of warrant officer.

Appendices

Text of HM King George VI's formal approval of the CGM

George the Sixth, by the Grace of God, of Great Britain, Ireland and the British Dominions beyond the Seas, King, Defender of the Faith, Emperor of India; To all to whom these Present shall come. Greeting.

Whereas Her late Majesty Queen Victoria was graciously pleased by Her Order-in-Council dated 7th July, 1874, to establish a medal designated the Conspicuous Gallantry Medal to members of Our Military and Air Forces for acts of conspicuous gallantry whilst flying in active operations against the enemy.

And whereas We deem it expedient to provide for the award of the Conspicuous Gallantry Medal to members of Our Military and Air Forces for acts of conspicuous gallantry whilst flying in active operations against the enemy.

Now therefore We do by these Presents for Us, Our Heirs and Successors, ordain that the following regulations shall govern the award of the said medal to members of Our Military and Air Forces:–

1. *Description.* – The Conspicuous Gallantry Medal shall be as described in the above-mentioned Order-in-Council, that is to say, it shall be silver and shall bear on the obverse the Royal Effigy and on the reverse the words 'For conspicuous gallantry' encircled by a wreath surmounted by a crown.

2. *Ribbon.* – The medal shall be worn on the left breast pendant from a ribbon one inch and a quarter in width which shall be in colour light blue with dark blue marginal stripes one eighth of an inch in width.

3. *Abbreviated title.* – The award of the medal shall entitle the recipient to have the initials C.G.M. appended to his name.

4. *Eligibility.* –

(1) The medal shall be granted to such persons as shall be recommended to Us by or through Our Secretary of State for air (or, in the case of any of Our Dominions the Government whereof shall so desire, the appropriate Minister of State for the said Dominion) for acts of conspicuous gallantry whilst flying in active operations against the enemy.

(2) The following shall be eligible for the medal:–

(a) warrant officers, non-commissioned officers and men of any Military or Air Force raised in Our United Kingdom of Great Britain and Northern Ireland, Our Indian Empire, Burma, any of Our Colonies or a territory under Our protection; or within any other part of Our Dominions Our Government whereof shall so desire or within any territory under Our protection administered by Us in such Government.

(b) foreign personnel, of ranks equivalent to those above-mentioned, who have been associated in operations with any of the aforesaid Military or Air Forces.

5. *Bars.* – When an individual who has been awarded the medal shall again be recommended to Us by or through Our Secretary of State for Air (or, in the case of any of Our Dominions the Government whereof shall so desire, the appropriate Minister of State for the said Dominion) for further conspicuous gallantry, he shall be awarded a bar to be attached to the ribbon by which the medal is suspended, and for every additional such recommendation an additional bar may be awarded. For every bar awarded a small silver rose shall be added to the ribbon when worn alone.

6. *Miniatures.* – Reproductions of the medal in miniature, which may be worn on certain occasions by those to whom the medal is awarded, shall be approximately half the size of the medal and sealed patterns of the miniature model shall be deposited and kept in the Central Chancery of Our Orders of Knighthood.

7. *Gazettement and registration.* – The names of those upon whom We may be pleased to confer the medal shall be published in the *London Gazette* and a register thereof shall be kept in the Office of Our Secretary of State for Air.

. 8. *Forfeiture and restoration.* – (1) It shall be competent for Us, Our Heirs and Successors by an Order under Our Sign Manual and on a recommendation by or through Our Secretary of State for Air (or, in the case of any of Our Dominions the Government whereof shall so desire, the appropriate Minister of State for the said Dominion) to cancel and annul the award to any person of the medal and thereupon the name of such person in the register shall be erased; but We, Our Heirs and Successors shall at all times have power to restore any medal which may have been forfeited when such recommendation shall have been withdrawn. (2) The forfeiture shall involve the cessation of any gratuity or pension in lieu thereof to which the possession of the medal might entitle the recipient, but no such forfeiture shall extend to any sum of money which has already been paid. (3) When a forfeited medal shall have been restored, any gratuity or pension in lieu thereof which attaches to it shall also be restored. (4) A notice of forfeiture and of restoration shall in every case be published in the *London Gazette.*

9. *Annulment, etc. of regulations.* – We reserve to Ourself, Our Heirs and Successors full power of annulling, altering, abrogating, augmenting, interpreting or dispensing with these regulations or any part thereof by a notification under Our Royal Sign Manual.

Given at Our Court of Saint James this
10th day of November, 1942 in the sixth
year of Our Reign.

By His Majesty's Command.

Recipients of the Conspicuous Gallantry Medal

Name	Sqdn	Aircraft	Duty	Date*	Gazetted
F/Sgt L.B. Wallace RNZAF	83	Lancaster	WOP	21/22 Dec 42	16 Feb 43
F/Sgt G. Ashplant	166	Wellington	Pilot	13/14 Feb 43	30 Mar
Sgt G.F. Dove DFM	101	Lancaster	AG	14/15 Feb	23 Mar
Sgt I.H. Hazard	101	Lancaster	Pilot	14/15 Feb	23 Mar
Sgt W.E. Williams	101	Lancaster	Nav	14/15 Feb	23 Mar
Sgt J.F. Bain	101	Lancaster	F/Eng	14/15 Feb	23 Mar
Sgt L. Airey	101	Lancaster	AG	14/15 Feb	23 Mar
Sgt E.W. Tickler	49	Lancaster	Pilot	27 Feb	2 Apl
F/Sgt C.C. Corder	248	Beaufighter	Nav	10 Mar	13 Apl
W/O H.F. Mc Taylor	156	Lancaster	Pilot	–	20 Apl
Sgt A.F. Elcoate DFM	156	Lancaster	WOP/AG	–	20 Apl
Sgt R.K. Hewitt DFM	61	Lancaster	WOP	10/11 Mar	14 May
F/Sgt G.F. Keen DFM	427	Wellington	WOP	12 Mar	23 Apl
Sgt L.F. Williamson	428	Wellington	Pilot	8/9 April	18 May
Sgt E.F. Hicks	466	Wellington	Pilot	14/15 Apl	14 May
F/Sgt R.S. Hogg	49	Lancaster	WOP/AG	–	14 May
Sgt J.P. McGarry	70	Wellington	Nav	12/13 Apl	4 Jun
Sgt T.P. Petrie	70	Wellington	Pilot	12/13 Apl	4 Jun
Sgt A.F. Blackwell RAAF	500	Hudson	Nav	23 Apl	1 Jun
Sgt G. Downton RAAF	1437	Baltimore	AG	21 Apl	23 Jul
W/O H. Vertican DFC	462	Halifax	Pilot	6/7 May	6 Jul
Sgt J.S. Powell	224	Liberator	Pilot	16 May	11 Jun
F/Sgt K.W. Brown RCAF	617	Lancaster	Pilot	16/17 May	28 May
F/Sgt W.C. Townsend DFM	617	Lancaster	Pilot	16/17 May	28 May
Sgt S.N. Sloan	431	Wellington	BA	23/24 May	11 Jun
F/Sgt I.W. Preece	106	Lancaster	AG	–	11 Jun
W/O B.W. Clayton	51	Lancaster	Pilot	–	11 Jun
Sgt N.F. Williams DFM RAAF	35	Lancaster	AG	11/12 Jun	6 Jul
W/O D.C.C.B. Busby	156	Lancaster	Pilot	–	9 Jul
F/Sgt F.E. Mathers RAAF	77	Halifax	Pilot	23/23 Jun	16 Jul
Sgt C.J.M. Wilkie	50	Lancaster	Pilot	28/29 Jun	23 Jul
Sgt E.T.G. Hall	115	Lancaster	AG	28/29 Jun	23 Jul
W/O M.G. Clynes	431	Halifax	AG	–	13 Aug
F/Sgt A.W.J. Larden RCAF	218	Stirling	BA	12/13 Aug	24 Sep
Sgt G.W. Oliver RAAF	467	Lancaster	AG	17/18 Aug	7 Sep
F/Sgt D. Rees RAAF	460	Lancaster	Pilot	17/18 Aug	24 Sep
Sgt J.C. Bailey RCAF	622	Stirling	BA	23/24 Aug	17 Sep
Sgt B.G. Bennett	623	Stirling	WOP	23/24 Aug	17 Sep

Name	Sqdn	Aircraft	Duty	Date*	Gazetted
F/Sgt O.H. White RNZAF	75	Stirling	Pilot	23/24 Aug	24 Sep
F/Sgt R.J. Foss	224	Liberator	Pilot	2 Sep	29 Oct
Sgt O.N. Jones	90	Stirling	F/Eng	22/23 Sep	19 Oct
W/O A.J.S. Walker	101	Lancaster	Pilot	27/28 Sep	2 Nov
Sgt S. Mayer	101	Lancaster	F/Eng	27/28 Sep	2 Nov
F/Sgt J.V. Russell RCAF	15	Stirling	Pilot	3/4 Oct	29 Oct
Sgt W.H. Cardy RCAF	427	Halifax	F/Eng	3/4 Oct	9 Nov
F/Sgt E.E. De Joux DFM	102	Halifax	AG	–	12 Nov
Sgt A.H. Cowham	57	Lancaster	AG	18/19 Oct	19 Nov
W/O C.E. White	100	Lancaster	Pilot	20/21 Oct	16 Nov
F/Sgt F.J. Stuart	426	Lancaster	Pilot	20/21 Oct	19 Nov
Sgt T.E. Bisby	10	Halifax	WOP	3/4 Nov	3 Dec
Sgt J.W. Norris	61	Lancaster	F/Eng	3/4 Nov	3 Dec
Sgt G.W. Meadows RCAF	166	Lancaster	AG	26/27 Nov	24 Dec
W/O E.S. Ellis	625	Lancaster	Pilot	2/3 Dec	24 Dec
W/O J.M. Alexander	7	Lancaster	Nav	–	18 Jan 44
W/O R. Haywood	7	Lancaster	WOP	–	18 Jan
W/O R.J. Meek RCAF	626	Lancaster	Nav	30/31 Jan 44	22 Feb
F/Sgt G.C.C. Smith RAAF	156	Lancaster	AG	15/16 Feb	7 Mar
F/Sgt C.R. Green RAAF	100 RAAF	Beaufort	Pilot	–	10 Mar
Sgt B.C. Wright	166	Lancaster	F/Eng	19/20 Feb	17 Mar
W/O G.W. Brook	550	Lancaster	Pilot	–	31 Mar
F/Sgt H.A. Donaldson	199	Stirling	WOP	5 Mar	31 Mar
Sgt L. Chapman	61	Lancaster	WOP	30/31 Mar	9 May
F/Sgt T.E. White	357	–	P.I.	March	2 Jun
Sgt E.D. Durrans	90	Stirling	WOP	10/11 Apl	9 May
W/O W.G. Bickley	617	Lancaster	AG	–	26 May
F/Sgt M.L. Langley RNZAF	489	Beaufighter	Pilot	6 May	16 Jun
F/Sgt J. Casson	250	Kittyhawk	Pilot	27 May	23 Jun
Sgt F.B. Dew	78	Halifax	F/Eng	27/28 May	7 Jul
Sgt P. Engbrecht RCAF	424	Halifax	AG	27/28 May	4 Aug
Sgt G.E.J. Steere RCAF	429	Halifax	F/Eng	7/8 Jun	21 Jul
W/O A.W. Hurse RAAF	75	Lancaster	BA	10/11 Jun	21 Jul
Sgt J.J. Reardon Jr USAAF	558 US	Marauder	BA	12 Jun	–
F/Sgt D.J. Moriarty RNZAF	75	Lancaster	Pilot	18 Jul	15 Sep
Sgt W.J. Bailey	78	Halifax	F/Eng	18/19 Jul	22 Sep
Sgt J. Pialucha	300	Lancaster	F/Eng	18 Jul	–
W/O H.A. Corbin	248	Mosquito	Pilot	14 Aug	17 Oct
F/Sgt S.S. Campbell	39	Beaufighter	Nav	–	24 Oct
F/Sgt R. Hartley	9	Lancaster	AG	–	24 Oct
F/Sgt R.B. Maxwell RCAF	428	Lancaster	Pilot	25/26 Aug	24 Oct
F/Sgt A.C. Cole	622	Lancaster	WOP	29/30 Aug	24 Oct
Sgt F.W. Cridge	166	Lancaster	Nav	23/24 Sep	10 Nov
Sgt E.W. Knight	432	Halifax	F/Eng	12 Oct	15 Dec
F/Sgt J.C. Cooke RCAF	103	Lancaster	Pilot	31 Oct	2 Jan 45
F/Sgt S.W. Walters	44	Lancaster	BA	1 Nov	2 Jan
Sgt D.J. Allen	467	Lancaster	AG	2 Nov	6 Mar

Name	Sqdn	Aircraft	Duty	Date*	Gazetted
F/Sgt F. Tomkins	180	Mitchell	WOP	3 Dec	23 Jan
F/Sgt R.P. Longley	218	Lancaster	WOP	6/7 Jan 45	13 Apl
F/Sgt W.E. Crabe RCAF	170	Lancaster	AG	1 Feb	20 Mar
F/Sgt T.W.D. Kelly	7	Lancaster	Nav	13/14 Feb	6 Apl
W/O V.A. Roe DFM	35	Lancaster	AG	–	13 Apl
F/Sgt J. Bettany	625	Lancaster	WOP	23/24 Feb	24 Apl
F/Sgt G.B. Ferguson RAAF	466	Halifax	AG	23/24 Feb	1 May
W/O A. Robb	405	Lancaster	AG	7/8 Mar	18 May
W/O K.J. Dennis	462	Halifax	WOP	13 Mar	13 Jul
F/Sgt J.M. Hall	180	Mitchell	AG	21 Mar	17 Apl
Sgt J.G. Wheeler	101	Lancaster	F/Eng	23 Mar	1 Jun
W/O L.W. Turnbull (Army)	–	–	Pilot	24 Mar	16 Aug
Sgt G.W. Simpson	463	Lancaster	F/Eng	25/26 Apl	3 Aug
F/Sgt D. Evans	250	Kittyhawk	Pilot	30 Apl	20 Jul
F/Sgt D.T. Jones	62	Dakota	Pilot	5 May	3 Aug
W/O L.E. Gosling	617	Lancaster	OBS	–	21 Sep
W/O A. Penrose DFC*	157	Mosquito	Nav	–	21 Sep
W/O S.C. Hopkins	214	Fortress	AG	–	26 Oct
W/O S. Nuttall DFM	35	Lancaster	AG	–	26 Oct
W/O H. Scott	223	Fortress	AG	–	26 Oct
W/O S.J. Tregunno	51	Halifax	AG	–	26 Oct
F/Sgt S.J.H. Andrew DFM	35	Lancaster	AG	–	26 Oct
F/Sgt A.H. Jefferies	550	Lancaster	Pilot	–	21 Dec
Sgt P.A. Hilton	35	Halifax	Pilot	2/3 Jun 42	29 Mar 45
Cpl J.D. Couglan RAAF	9 RAAF	Helicopter		–	10 Dec 68

* If awarded for a specific action.
Nav   – Navigator
WOP – wireless operator
BA   – bomb aimer
AG   – air gunner
OBS – observer
PI   – parachuter instructor
F/Eng – flight engineer

Conspicious Gallantry Awards by crew category

Pilots	36
Air Gunners	26
WOPS	16
F/Engineers	13
Navigators	10
Bomb aimers	6
Observers	1
Parachute Inst	1
Winchman	1
	110

CGMs awarded

Royal Air Force	80
Royal Australian Air Force	10
Royal Canadian Air Force	12
Royal New Zealand Air Force	4
Polish Air Force	1
US Army Air Force	1
Glider Pilot Regt	1
Royal Australian Air Force – Post War	1
	110

Other Awards to Holders of the CGM

Sergeant G.F. Dove	DFM	18 April 1941
Sergeant A.F. Elcoate	DFM	18 April 1941
Flight Sergeant E.E. De Joux	DFM	9 January 1942
Flight Sergeant G.F. Keen	DFM	30 January 1942
Sergeant R.K. Hewitt	DFM	13 March 1943
Flight Sergeant W.C. Townsend	DFM	14 May 1943
Sergeant N.F. Williams	DFM	24 November 1942
Sergeant N.F. Williams	bar	18 May 1943
Warrant Officer B.W. Clayton	DFC	15 April 1943
Flight Lieutenant B.W. Clayton	DSO	26 September 1944
Warrant Officer H. Vertican	DFC	14 May 1943
Pilot Officer H.F. McTaylor	DFC	15 June 1943
Pilot Officer E.F. Hicks	DFC	16 November 1943
Pilot Officer E.H. Hicks	MID	1 January 1945
Sergeant V.A. Roe	DFM	13 June 1944
Warrant Officer A. Penrose	DFC	20 October 1944
Warrant Officer A. Penrose	bar	9 March 1945
Flight Lieutenant E.S. Ellis	DFC	6 June 1944
Sergeant S.J.H. Andrew	DFM	8 December 1944
Sergeant S. Nuttall	DFM	12 December 1944
Sergeant T.W.D. Kelly	DFM	17 April 1945
Flying Officer C.R. Green	MID	4 May 1945
Flying Officer R.B. Maxwell	DFC	25 September 1945

CGMs awarded by aircraft types

Avros Lancaster	59
Handley Page Halifax	15
Short Stirling	8
Vickers Wellington	7
Bristol Beaufighter	3
Curtis P40 Kittyhawk	2
DeHavilland Mosquito	2
North American Mitchell	2
Consolidated Liberator	2
Boeing Fortress	2
Bristol Beaufort	1
Douglas Dakota	1
Martin Marauder	1
Lockheed Hudson	1
Martin Baltimore	1
Helicopter	1
Glider	1

109 + one parachute instructor = 110

CGMs awarded by squadrons

7 Squadron	3	156 Squadron	4
9 Squadron	1	157 Squadron	1
10 Squadron	1	166 Squadron	4
10 Squadron RAAF	1	170 Squadron	1
15 Squadron	1	180 Squadron	2
35 Squadron	5	199 Squadron	1
39 Squadron	1	214 Squadron	1
44 Squadron	1	218 Squadron	2
49 Squadron	2	223 Squadron	1
50 Squadron	1	224 Squadron	2
51 Squadron	2	248 Squadron	2
57 Squadron	1	250 Squadron	2
61 Squadron	3	300 Polish Squadron	1
62 Squadron	1	357 Squadron	1
70 Squadron	2	405 Squadron RCAF	1
75 NZ Squadron	3	424 Squadron RCAF	1
77 Squadron	1	426 Squadron RCAF	1
78 Squadron	2	427 Squadron RCAF	2
83 Squadron	1	428 Squadron RCAF	2
90 Squadron	2	429 Squadron RCAF	1
100 Squadron	1	431 Squadron RCAF	2
100 Squadron RAAF	1	432 Squadron RCAF	1
101 Squadron	8	460 Squadron RAAF	1
102 Squadron	1	462 Squadron RAAF	2
103 Squadron	1	463 Squadron RAAF	1
106 Squadron	1	466 Squadron RAAF	2
115 Squadron	1	467 Squadron RAAF	2

489 Squadron RNZAF 1
500 Squadron   1
550 Squadron RCAF   2
558 Squadron USAAF 1
617 Squadron   4
622 Squadron   2
623 Squadron   1
625 Squadron   2
626 Squadron   1
1437 Flight   1
                  ———

109 + one Glider Pilot Regt

Index